QGIS 3 TUTORIAL FOR
BEGINNERS #1

GETTING STARTED

QGIS Tutorial and
Video Course

Ian Allan

Title Copyright © 2019 by Ian Allan. All Rights Reserved.

All rights reserved. No part of this book may be reproduced in any form or by any electronic or mechanical means including information storage and retrieval systems, without permission in writing from the author. The only exception is by a reviewer, who may quote short excerpts in a review.

Cover designed by

nitty gritty graphics

nittygrittygraphics.com.au

Ian Allan

Visit my website at gis-university.com

First Printing: February 2019
Minor update: January 2020

Geocode Mapping and Analysis pl

ISBN: 9781797834238

The problem with maps in a GIS is that its...
Garbage in, Garbage Out.
Or was that...
Garbage in, Gospel Out?

IAN ALLAN

Other QGIS Tutorials by Ian Allan 1
About Ian Allan 2
Introduction 5
   GIS Data Concepts 5
      Spatial Data (Maps) 7
      Attribute Data (Things about the maps) 9
      Explicit Data Relationships 9
      Implicit Data Relationships 11
      Combined Spatial and Database Query 11
      Data Currency and Data Consistency 11
   Map Overlay 12
   The basics of Map Scale as it applies to GIS 14
   Some tangible advice on map scale in GIS 16
Setting up 17
   This tutorial uses QGIS 3 for Microsoft Windows 17
   How to Download QGIS 18
   How to Install QGIS 21
   How to Uninstall Quantum GIS (if you ever need to) . 26
   How to download the Sample Dataset 26
   How to Launch QGIS 29
   How to check for Updates 30
   How to Make Your QGIS Interface Look Like Mine 30
      Panel Options 32
      Toolbars 33
The QGIS Desktop 35
   The Map Layers Window and the Map Window 37
      The Layers Window 38

- Layer Properties 39
- The Map Window 41
- **The Data Source Manager Toolbar 42**
  - Add a Vector Map (cadastre) 42
  - Add a Raster Map (air photo) 44
- **The Map Navigation Toolbar 49**
- **The Attributes Toolbar 50**
  - Identify Features Button 51
  - The Geographical Selection Button 52
  - Selecting Some or All Features in Your Map 54
  - Open Attribute Table Button 58
  - Measurement Tools 64
- **Bottom Status Bar 68**
- **Visual Interpretation of the Vector Property Map and Air Photo Map 69**

Shading a Map 75

- **Background: GIS Tables 77**
- **Background: Map Generalization 79**
- **Background: Time Series Mapping 81**
- **How to Shade a Map using Categorized Data 83**
- **How to Validate Your Map 93**

Conclusion 99

Coupon for the Companion Video Course 101

Glossary of Terms 103

OTHER QGIS TUTORIALS BY IAN ALLAN

All my QGIS tutorials are aimed at researchers, college students and professionals who just want to learn the essentials of a mapping tool. Every text is paired with a companion video course hosted on Udemy.com. I teach you, step-by-step, the essential elements of a QGIS task.

GIS 3 for Beginners #4: Learn to Geocode

Learn how to map addresses in your spreadsheets.

A geocode is a text description that can be related to something on a map. For example, addresses, suburb names and postal codes.

Available on Amazon.com

https://www.amazon.com/Learn-Geocode-Using-QGIS-Beginners-ebook/dp/B07K65VSR5/

ABOUT IAN ALLAN

I have authored and co-authored fifteen peer reviewed publications. I have worked professionally as a GIS researcher, taught GIS to thousands of students, and worked as a GIS consultant on projects as diverse as the following…

- **United Nations:** Post tsunami strategic planning in Banda Ache
- **Australian Federal Government:** National Broadband strategic assessment.
- **Victoria Australia's Department of Premier and Cabinet:** Housing affordability modelling.
- **Local Government**: Environmental sustainability modelling for planners.
- **Water industry:** Buried water pipe condition modelling and ease-of-digging modelling.

Since the mid 1990's I've been a GIS researcher, teacher and consultant. Over 5000 students have enrolled in my Udemy GIS courses. Here's what some of my students say about my teaching style…

Caesar says: It was a very good class, just what I needed to get familiar with QGIS.

Carina says: "Perfect course to whom have never used…QGIS! Very detailed on the explanations and really generous additional materials to study."

Brian says: "It was very thorough and comprehensive… Ian also covered aspects of GIS Analysis - which increases your learning and appreciation of the capabilities of how powerful the GIS tool can be"

Umar says: "Ian is vast and knowledgeable in what he teaches. I would do another course by Ian if he offered it"

Nathan says: "Really great introductory-level course! The examples were simple to follow, but also very useful. All the work can be finished while watching the videos, there was no extra work to be done without guidance. I really look forward to the next course by this instructor. - Cheers and well done!"

Boojhawon says: "Simple and very clear lectures with the minimum basics/backgrounds to get a taste of what awaits us further and also making us think of what we can do more."

SRIJON says: "I have experience in using ArcGIS. I needed the QGIS taste for which I took this. If anybody who has little experience in GIS will love the 'Techniques and Tips' parts, which are very clearly and elaborately described, which will greatly help to be a GIS interpreter. And I love the line "Don't just be a tourist." It should be the tricky line between a GIS analyst and a technician. The course will help me in my future works."

IAN ALLAN

INTRODUCTION

GIS can seem complex when you first start out and it is common to be overwhelmed. In the absence of someone holding your hand, there is a steep learning curve as you come to grips with the three facets of GIS – maps, databases and software.

Learning GIS need not be difficult. Few people, even power-users, use the full breadth of it. Most people only use the basics (zooming and panning, shading, overlaying, digitising, and presenting maps). Power users build on the basics and then tend to spend most of their time in a small area of one-or-another drop-down menu. That is why in this first tutorial I simply show you how to install QGIS, find your away around the interface, open maps, and then shade them.

Some of you will come to this tutorial with specific aims and so will skip to the sections that most interests you. However, if you do not have a background in geography or cartography, please read the two sections about the basics of map scale as it applies to GIS.

GIS Data Concepts

Why should we bother about desktop GIS systems such as QGIS when web based systems such as Google Maps have everything there right? Well, not quite! You see, Google maps and others like it do a fantastic job of bringing together a bunch of useful maps and other data. They allow you to do basic search and visual analysis. However, if you have your own GIS maps, it's a tricky task to bring them into Google maps, and even when they're there, there's no meaningful analytical functionality available to you. This is where QGIS comes into play. It does all that Google Maps does and much more. QGIS allows you to…

- Create maps as well as display them.
- Explore the relationships that are often "implicit" in spatial data.
- Undertake quantitative analysis as well as visual analysis.
- Create cartographic quality maps.

In this section I'm going to talk about the types of data that GISs use. With a GIS database in place, you can quantitatively explore relationships that are EXPLICIT in your data, and those that are IMPLICIT in your maps. Beware of data quality issues though!

Spatial Data (Maps)

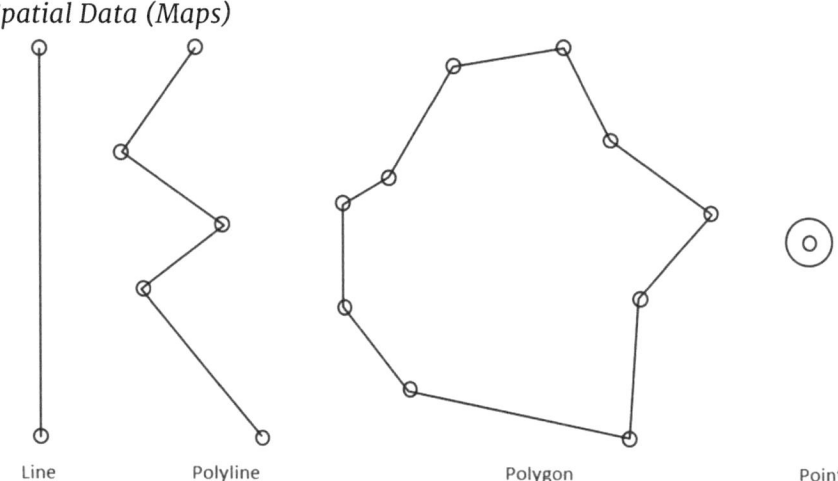

Figure 1: *The four GIS object types are Lines, Polylines, Polygons and Points. The small round circles are called nodes.*

Every "thing" on a GIS map (road, letter box, power pole, etc) is called an object. The four types of GIS objects are *Points, Lines, Polylines* and *Polygons*. These simple constructs are extremely powerful in a GIS because you can relate them to each other to answer geographical questions.

Each object type can be imagined to be one or more mouse clicks. These mouse-clicks are known as nodes and are shown as tiny circles in Figure 1.

- **Point (one mouse click):** You create a point by using your mouse to click once on your GIS map. A point might be a tree, power pole, fence post, rubbish bin, address point, etc

- **Line (two mouse clicks):** You create a line by using your mouse to click twice on your GIS map – once for the start point and once for the end point. A line might be a section of road, water pipe, etc.

- **Polyline (three or more mouse clicks):** You create a polyline by using your mouse to click in three or more places on your GIS map – once for the start point, as many mid points as you need, and once for the end point. A polyline might be a curved road, river, etc.

- **Polygon (four or more mouse clicks):** You create a polygon by using your mouse to click four or more places on your GIS map – once for the start point, as many mid points as you need, and once for the end point. The end point must be identical to the start point. A polygon might be a land parcel, sports field, planning zone boundary, census collection area, etc.

Each QGIS map can only contain one type of GIS object.

When you're first told about GIS points, lines and polygons, they sound to be very simple. Often they are not. For example…

- **Lines and Polylines:** If you digitize a river with the start-point always upstream and the end-point always downstream, then sophisticated GIS routines can work out the direction of water flow. These same data quality principles allow you to undertake network analysis of roads. Emergency services use such information to locate a fire or ambulance station within say five minutes drive of all the houses that its meant to service.

- **Polygons:** Polygons that join each other must share "identical" boundaries. This means that all mouse clicks describing the boundary between say, two blocks of land, must be in the same place. If boundaries overlap then this can cause all sorts of problems. For example…
    - Incorrect land size calculation.
    - Conflict between neighbours.

There are simple GIS map structures. There are also complex map structures that are "topological". Both are invisible to the GIS user. It is a good idea to have a broad understanding, both that they exist, and how they're structured.

For the *Shape* files that we're using in this tutorial, the boundary that two adjoining parcels of land share in the cadastre map is repeated for each parcel.

With more sophisticated *topological* map structures, there is only line describing the boundary of adjoining land parcels. That line contains identifying information for the block of land on the left side of the line, and the right side of the line.

Attribute Data (Things about the maps)

Table 1: The four GIS object types with examples of the types of data that might be attached to them

Object type	Mouse clicks	Possible attribute #1	Possible attribute #2
Point	One	Power pole	Wooden
Line	Two	Straight road	Smith Street
Polyline	Three or more	River	Poor condition
Polygon	Four or more, with the last being identical to the first	Land parcel	Owned by Ms Smith

In the absence of attribute data GIS maps are dumb. They may as well exist in a graphics package where roads could be coloured and weighted, rivers shaded blue, forests green, etc. Although the graphics map would be logical to the eye, you would not be able to query it. You could not ask questions like "how many acres of forest" or "where is Smithsville". To do that, attributes need to be attached to the objects on your GIS map. Examples of these are shown in Table 1.

Explicit Data Relationships

By EXPLICIT I mean a relational database design where two tables are related to each other via a KEY field. Customer ID, Ratepayer ID and Student ID are examples of key fields. In a well designed relational database, no data are stored more than once. The absence of duplicated information eliminates the possibility of conflicting data.

One way to think of a relational database is to imagine it is a Spread Sheet Workbook. The Workbook would be the database and each WorkSheet would be a table (Figure 2).

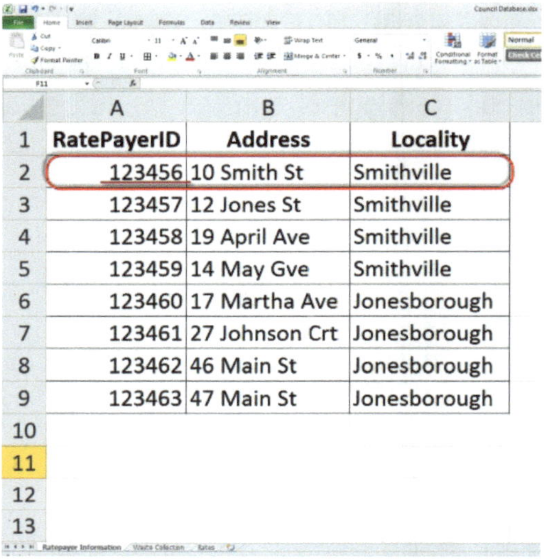

	A	B	C
1	RatePayerID	Address	Locality
2	123456	10 Smith St	Smithville
3	123457	12 Jones St	Smithville
4	123458	19 April Ave	Smithville
5	123459	14 May Gve	Smithville
6	123460	17 Martha Ave	Jonesborough
7	123461	27 Johnson Crt	Jonesborough
8	123462	46 Main St	Jonesborough
9	123463	47 Main St	Jonesborough

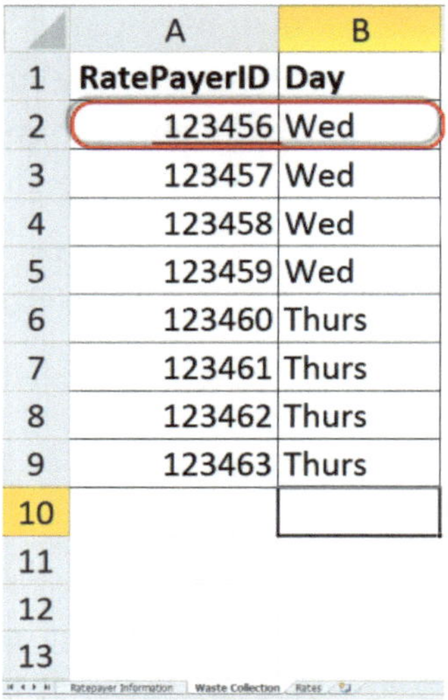

	A	B
1	RatePayerID	Day
2	123456	Wed
3	123457	Wed
4	123458	Wed
5	123459	Wed
6	123460	Thurs
7	123461	Thurs
8	123462	Thurs
9	123463	Thurs

Figure 2: Try thinking of a database as being a spreadsheet workbook. Each worksheet equates to a table that holds one type of information. In this example its a Local Government database with tables holding ratepayer address, what day their rubbish gets collected and their rates. The two tables can be related to each other by a Key Field called RatePayerID. Using a query along the lines of "look up the Ratepayer Information table and the Waste Collection table and tell me the address of each ratepayer and the day their rubbish gets picked up" would deliver the information that 10 Smith St Smithville has its rubbish collected every Wednesday.

Implicit Data Relationships

By IMPLICIT I mean that two or more things can be related implicitly based on their geography. Something might be NEXT TO something else (eg. a house next to a freeway), or something might be WITHIN something else (eg. a rare plant within a park, or a house within a postal code). An implicit spatial query of a real estate system might ask "how many houses were sold *within* each postal area last year"?

The exciting thing about making use of geographical relationships is that databases that previously could not be related to each other, can be. Imagine having two maps covering the same brownfield development...

- an old map that showed industries that in their day were toxic to the environment, and
- a new map that showed residential cadastre.

The new map could be overlayed onto the old map. Visually, this would *implicitly* show all houses that are built on land that was previously occupied by toxic industries.

Combined Spatial and Database Query

Combining explicit and implicit data queries can be phenomenally powerful. The *implicit* relationship between the toxic industry map and the residential map could be followed by an *explicit* interrogation of a rates database. Contamination warnings could then be sent to the affected households.

Data Currency and Data Consistency

The issues of data currency and data consistency are very important in GIS. Without paying close attention to these issues, you could unwittingly create maps that are wrong. GIS is only as good as the data it uses.

Data Currency

In the same way that a business needs to keep its accounts current, you also need to keep the data that lies behind your maps current. For example, a google map of restaurants that was created in the early 1990s is unlikely to still be correct unless it was regularly updated.

Data Consistency

It is important to maintain consistent standards in your database. Entries that appear logical to people are not always logical to a computer. For example if you enter "Cantonese Restaurant" for one establishment and "Chinese Restaurant" for another, then the computer will see these as two different types of restaurants, even though, to the lay-person, they're both Chinese restaurants. So be sure to adopt a standard and stick to it!

Map Overlay

Quantum GIS has its roots in cartographic analysis of the kind demonstrated by Ian McHarg in his 1967 book *Design with Nature*. This is one of the earliest examples of the strategic use of map overlay in planning. If you understand map overlay then you are a long way along the path to understanding GIS.

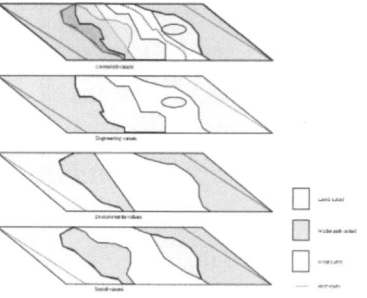

Figure 3: *A conceptual view of Ian McHarg's map overlay technique.*

Geographers have been using analogue GIS for many years, and only since the 1980s has high powered desktop computing brought about the routine use of computerized GIS. Traditionally, geographers have overlaid maps on top of each other on light tables and traced new thematic maps. The great difficulty with this has always been that they could only overlay a few maps onto each other at the time, and that the maps had to be of the same scale.

In the 1960s, Ian McHarg applied his map overlay technique to a site selection study for a proposed road in New York. He created a series of around twenty transparent thematic maps at the same scale and overlaid them. Some maps related to social values, some environmental values, and others to engineering values…

- The darkest shades depicted the areas that were least suitable for the road (they had the lowest soil and engineering values and the highest environmental values).

- The transparent shades depicted the areas that were most suited to having the road (they had the highest engineering values, worst social values, and the worst environmental values).

When he overlaid the maps, the transparent areas showed those areas that were most suitable for the road, and opaque areas those areas least suitable. Conceptually, this can be seen in Figure 3.

McHarg was hailed as a revolutionary and more than fifty years on, his book *Design With Nature* is still a prescribed university text. Although geographers had been overlaying maps on light tables for many years, Ian McHarg's approach was different because he took the time to get many themes over a large area into good condition so that they could be overlaid. The same principle applies to you when you use GIS. You will get the most value out of QGIS if you take the time to make your mapping the highest practical quality for the purpose you intend to use it. You do not need to use sophisticated GIS routines to be innovative or productive. In fact, the worst thing you could do is to undertake very sophisticated analyses using poor quality map data. The old saying is *garbage in, garbage out!*

The basics of Map Scale as it applies to GIS

Many GIS maps have their origins in paper maps. These paper maps were produced by cartographers to be read for a purpose, at a scale. For example...

- **Purpose:** Walking Maps highlight features important to walkers. Street Maps highlight features important to road users.
- **Scale:** A national scale Road Atlas shows only major roads. A local Street Directory shows every street.

Inexperienced GIS users sometimes fall into the trap of using maps in ways that their authors never intended. Once in GIS all maps become scale-less, so it is important that you understand the pedigree of the maps you're using, and whether they're appropriate for the purpose you're using them. And then with this combined knowledge, impose a scale onto any GIS maps you produce.

Here's an example of a bad understanding of scale. I once saw a GIS officer present their planning map to a seminar. The map they created, related a mix of 1:250,000 scale, 1:100,000 scale and 1:50,000 scale Rural Development Land Suitability maps to 1:10,000 scale semi-urban cadastre (land ownership). They provided this map to their planning department who then used it as a basis for imposing planning restrictions onto ratepayers. None of the Land Suitability maps were detailed enough to be used for that purpose!

You need to be aware of scale issues when you're using GIS some maps, but not others.

There are no scale issues for maps that have been purposely surveyed for use in GIS (eg. cadastre).

Scale does matter for thematic maps (eg. soils, vegetation, geology, etc). That is because the scale of a thematic map relates directly to the amount of effort that went into creating it. A map produced at the end of a week in the field will be of poor quality compared to a map produced by the same people over the same area following a year of fieldwork!

A mistake that people sometimes make is to enlarge a map on a photocopier in the hope of producing more detailed mapping. Sometimes it is okay to do this, but increased size does not equate to increased detail – no matter how many times you enlarge a world atlas on a photocopier, local streets do not appear!

The take-home message is that whenever possible, the maps you use for your projects should be at an appropriate scale for the project's purpose. Because the scale of a map relates to the scale of its collection, and not the scale of its display, whenever appropriate, on your printouts you should indicate the scale of the original mapping as well as your GIS's auto-generated scale bar.

Some tangible advice on map scale in GIS

- The GIS version of a 1:250,000 scale paper map printed/displayed at a scale of 1:10,000, is not a 1:10,000 scale map!

- Maps created by licensed surveyors using modern surveying equipment are legitimately scaleless in a GIS. These are (but not always) maps such as cadastre (land ownership), roads, administrative boundaries (suburbs, electoral boundaries, etc).

- It's better to use a map created from multiple maps that have vastly different scales at small scale rather than large scale.

- Maps of interpretations require thoughtful use. A map created from multiple maps that were created using different standards is often wrong!

    o **Demographic maps:** Even if demographic boundaries are accurate, when the data attached to them have been created by different organizations using different standards and data definitions, they are most likely incompatible. For example, the definition of what it means to be unemployed often changes from census to census.

    o **Environmental maps:** These might have been created by different organizations, using different standards, old fashioned surveying equipment, or different understandings of the phenomena being mapped. For example, I've used soils maps from the 1940s that were "thematically correct", but when compared to surveyor-produced contour maps, were displaced by as much as two kilometres.

SETTING UP

Setting up for the Tutorial

Downloading open source software off the Internet, installing it, and using it for the first time can be intimidating. It does not need to be that way. What I show you in this section is…

Where to find Quantum GIS on the Internet.

How to download and install QGIS.

QGIS is a fully functional GIS. The major difference between QGIS and commercial GISs is that commercial GISs are mostly supported under commercial support arrangements, while QGIS is both written and supported by a "community". Neither support model is perfect.

The concepts you will learn in this tutorial are directly applicable to all the major geographical information systems.

This tutorial uses QGIS 3 for Microsoft Windows

I teach this QGIS tutorial using QGIS for Microsoft Windows. QGIS also runs on Linux, MacOS X and Android. QGIS should look very similar on every platform. However, **I cannot support platforms other than MS Windows**.

Having said that, you should search the Q&A area for solutions to version specific problems that occur from time-to-time.

How to Download QGIS

Let's download and install QGIS. Follow the instructions in Figure 4 thru Figure 8 <u>closely</u> and watch the Udemy video. I'd be surprised if you had any problems.

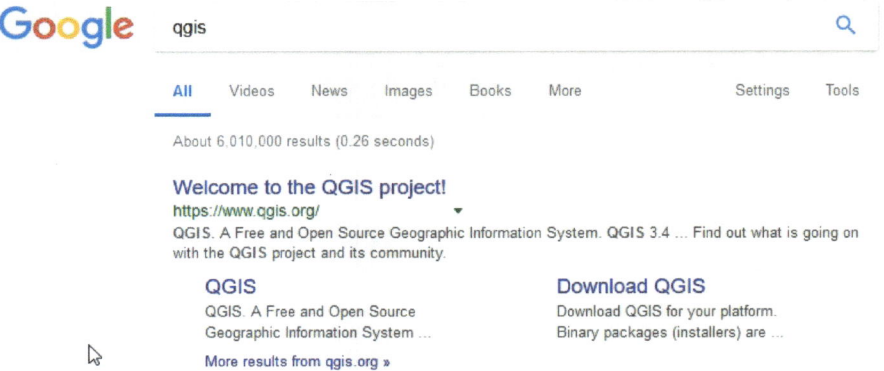

Figure 4: Begin by using your favorite search engine to find the QGIS site (qgis.org).

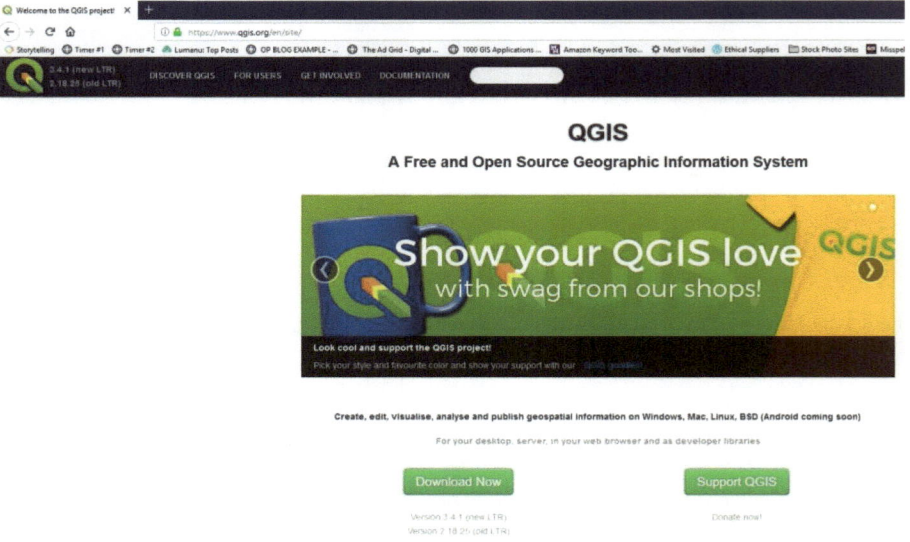

Figure 5: The QGIS welcome screen changes all the time. It should look something like this. You need to click on the "Download Now button.

QGIS 3 TUTORIAL FOR BEGINNERS #1: GETTING STARTED

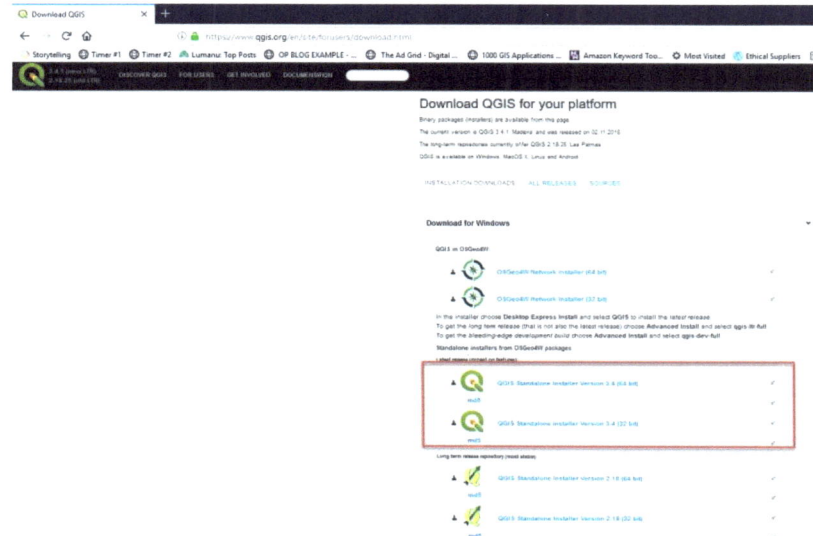

Figure 6: From the QGIS page choose to install the "Standalone Installer". I have found the 32 bit version to be a bit more stable than the 64 bit version. You only need the 64 bit version if you have large datasets, or if the 32 bit version is slow or otherwise not coping.

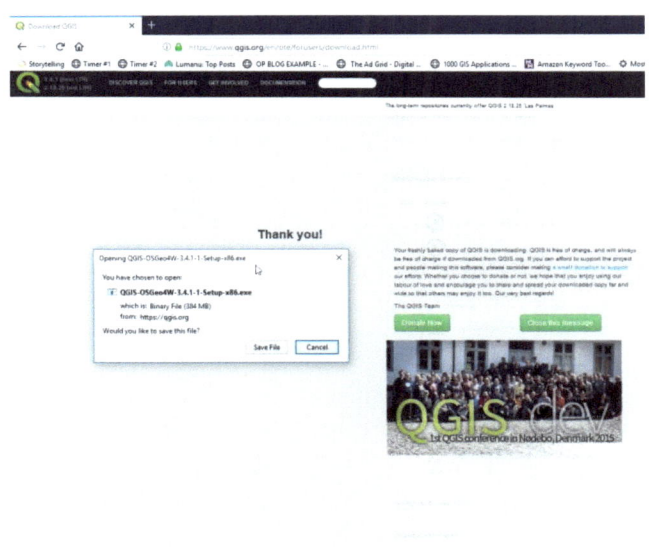

Figure 7: Nearly there. Next, you'll get an option to "Save File". Click on the button and the download will begin. It's a big file so be patient. If you have a slow internet connection this may be a good time to take a break.

Figure 8: Check your downloads folder for the installation file. The 32 bit version is around 400mb. If the file you download is much larger or smaller than this, or the name is quite different, then you should double-check that you're downloading the correct file. Be sure that you've been downloading the "Standalone Installer"

How to Install QGIS

Installing QGIS is as easy as double-clicking the installation file you downloaded and then accepting all the recommended defaults as the installation proceeds. Figure 9 thru Figure 17 show you what to expect.

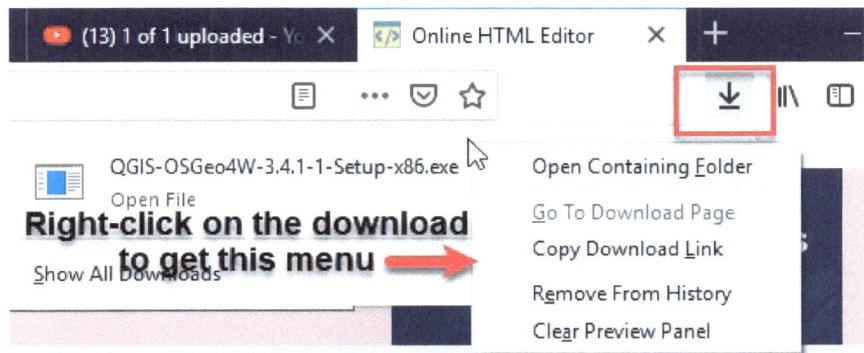

Figure 9: Once the file is downloaded you should open the folder that the file was saved to. In FireFox, access the Downloads dialog via the down-arrow at the top right corner of your browser. When in the downloads dialog, right-click on the file. An option to "Open Containing Folder" will appear. Click on that.

The way you find the downloaded file is similar in most web browsers. If you still can't find Download folder, type your problem as a question into Google. For example, Google "where is the Chrome downloads folder".

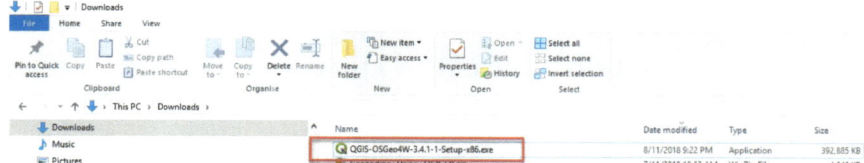

Figure 10: In the Downloads folder double-click on the QGIS setup file with your left mouse button and the installation program will launch.

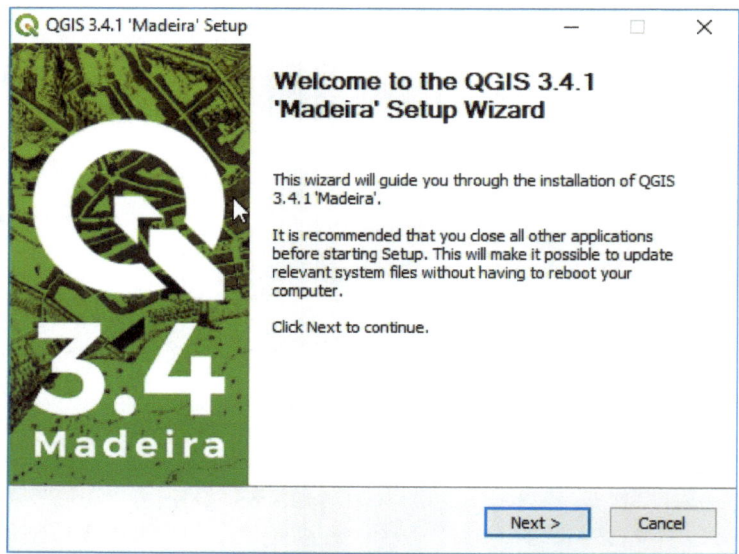

Figure 11: Now just following the on-screen instructions. Click the Next > button

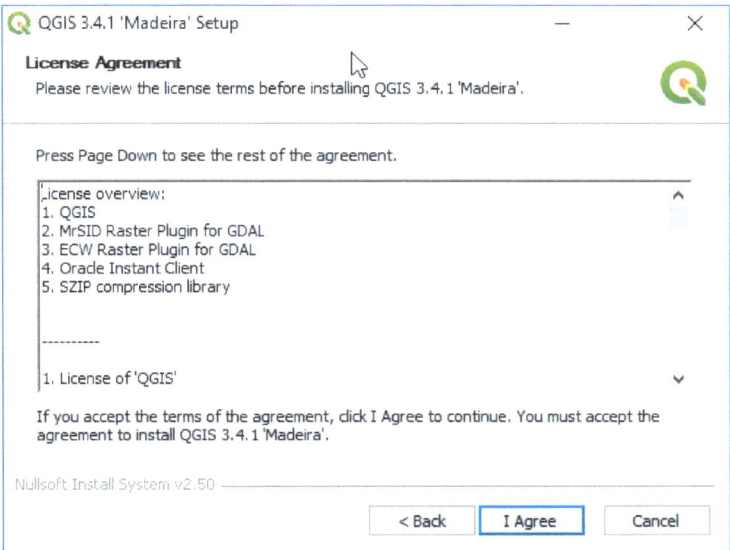

Figure 12: Now you're presented with the option to either accept or not accept the QGIS license agreement. If you choose not to, you won't be able to install the software. I recommend that you click the "I Agree" button and just get on with the installation!

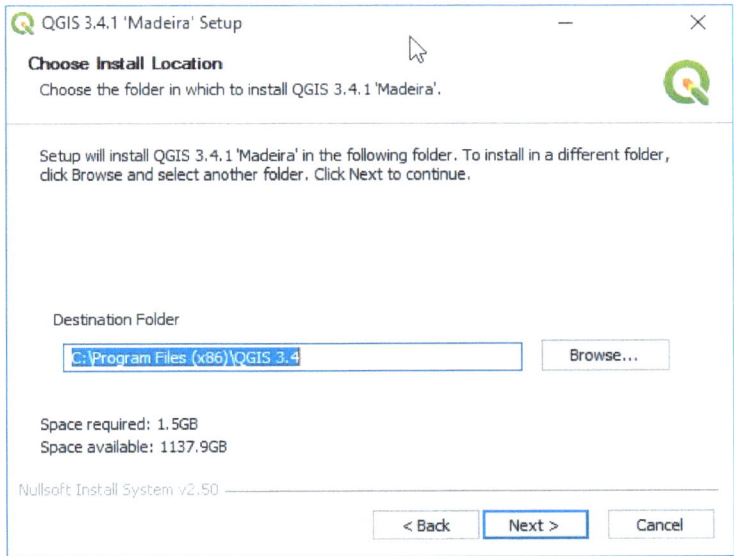

Figure 13: Unless you have a really good reason for doing otherwise, you should allow QGIS to install itself in the default directory path.

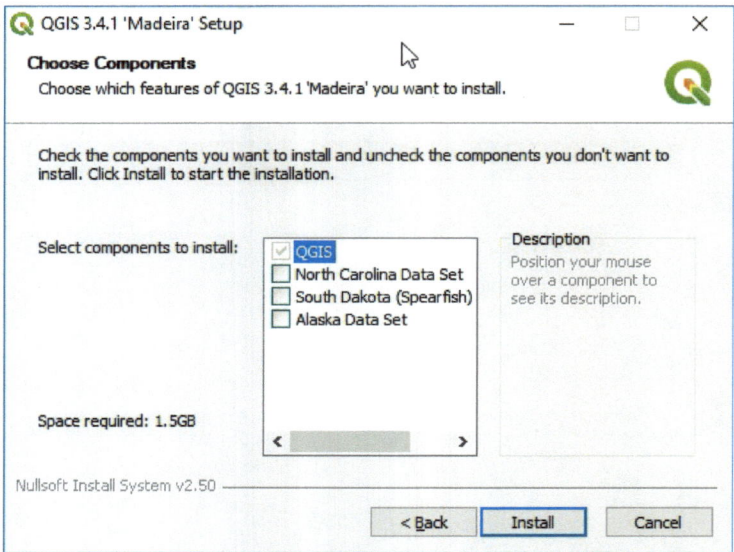

Figure 14: You can choose to install the sample datasets if you want, but it's not necessary because I'll be supplying you with the GIS datasets so you can follow along with the videos. So, just click the Install button.

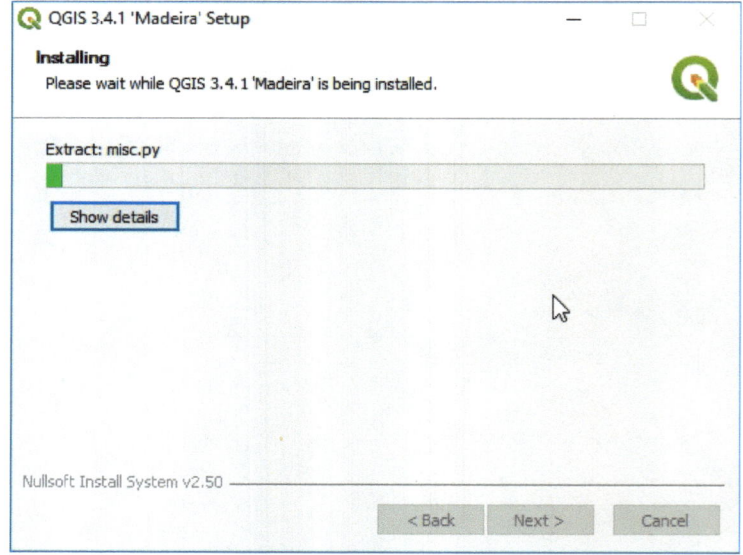

Figure 15: QGIS can take a little while to install. Take a break if you need one.

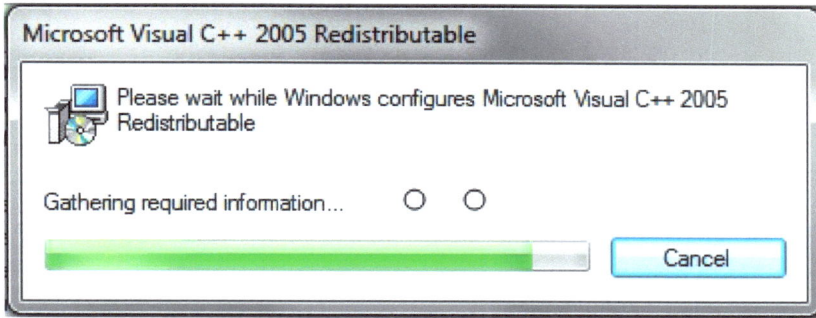

Figure 16: Its possible that QGIS will install other software that it needs. Don't worry if you see boxes like this one pop-up.

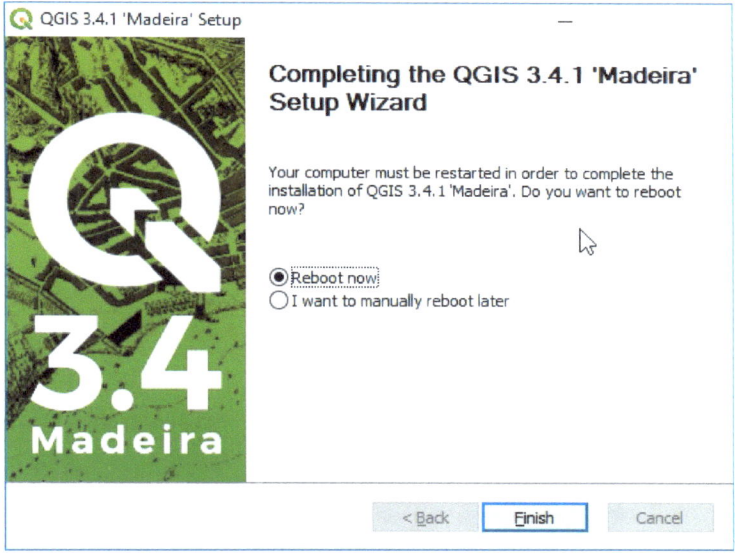

Figure 17: Voila! You're finished! Reboot and QGIS 3 is installed! Congratulations!

How to Uninstall Quantum GIS (if you ever need to)

If ever you need to uninstall QGIS, you do that from the Apps & Features area in the Windows Control Panel (Figure 18).

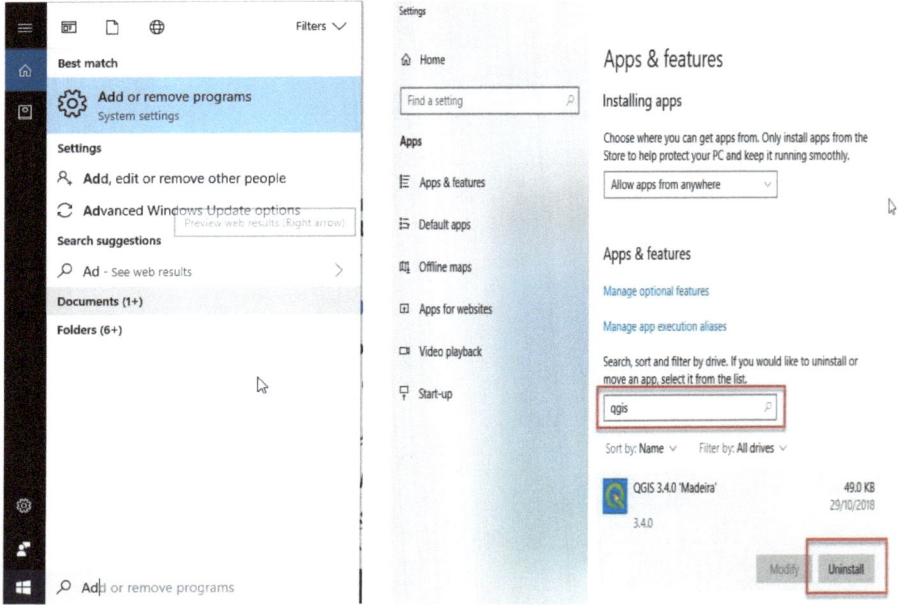

Figure 18: To uninstall QGIS, type "Add or Remove Programs" in the windows search area. In the Apps and Features dialog, type QGIS into the search box and then click on the Uninstall button. From there just click on the Uninstall button and follow the prompts.

How to download the Sample Dataset

You download the sample dataset from the resources area of the Udemy lecture titled…

"QGIS 3.4: Download the practice activity dataset here".

The resources area appears when you hover your mouse cursor in the top-left corner of the screen. Click on the zip file link it will download.

Students who purchased this tutorial from *Amazon* will find a coupon code to access the videos and dataset under the *Coupon for the Companion Video Course* heading on page 101 of this book.

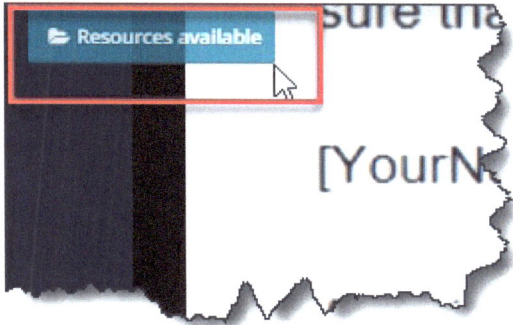

Figure 19: Download QGIS Orientation.zip from the Resources Available area on the top left of the "QGIS 3.4: Download the practice activity dataset here" lecture.

It will be much easier for you to follow along with this tutorial if you place the downloaded **QGIS Orientation.zip** file into a folder called **QGIS for Beginners** on your desktop. Then when you unzip the file, the contents should unzip into a folder called **QGIS Orientation** and you'll end up with a directory structure that's the same as in the Udemy videos...

..\Desktop\QGIS for Beginners\QGIS Orientation

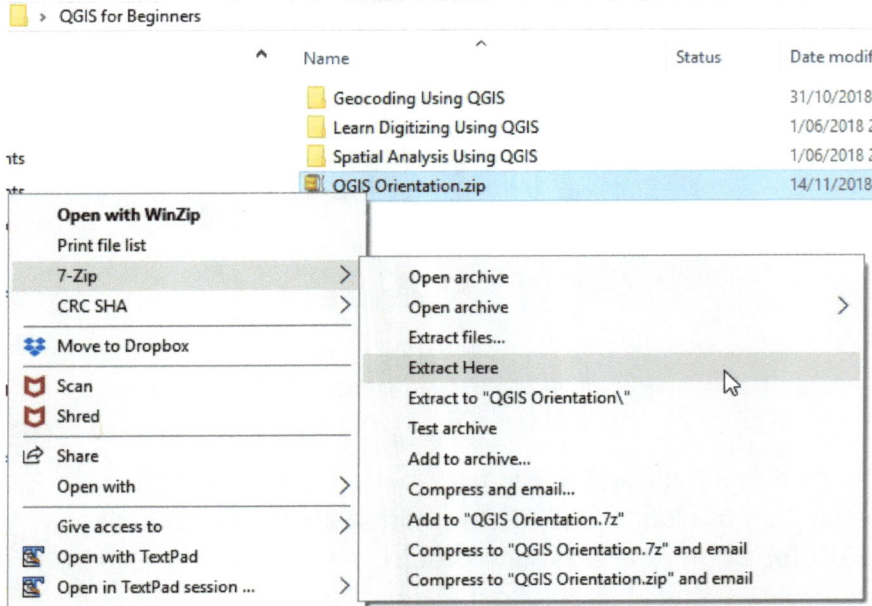

Figure 20: This is what your folder structure should look like. The other folders in the screen capture relate to my other Udemy QGIS tutorials.

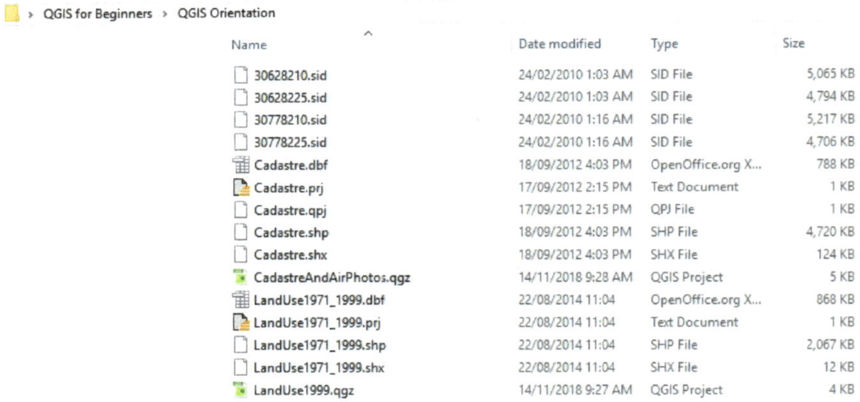

Figure 21: This is what your folder should look like when the datasets have been extracted. I used 7-Zip, but you could use WinZip if you wanted.

At this point I would like to very gratefully acknowledge the Office of Geographic Information (MassGIS), Commonwealth of Massachusetts, Information Technology Division. They have compiled and made available the GIS data used in this tutorial.

How to Launch QGIS

The three main ways to launch QGIS are shown in Figure 22. They are also described in the dot points below...

- **Desktop Icon:** Move your mouse pointer to the icon and then double-click using the left mouse button.

- **Start Menu (recently used):** If QGIS has been used recently, it's likely that it will be shown in the recently used programs section. Move your mouse to it and single-click using the left mouse button.

- **Start Menu (All Programs):** You can find QGIS in the All Programs area from within the Windows Start menu. Move your mouse to it and single-click using the left mouse button.

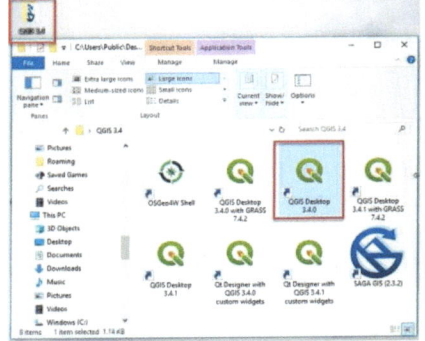 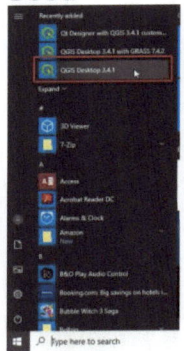 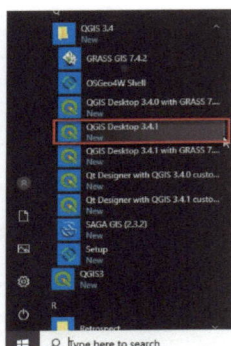

Figure 22: The three ways to launch Quantum GIS.

How to check for Updates

You can check for updates from the help menu by clicking on the Check QGIS version option. The menus and dialog boxes you can expect to see are shown in Figure 23.

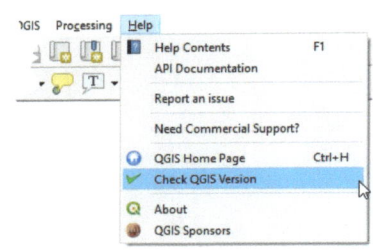 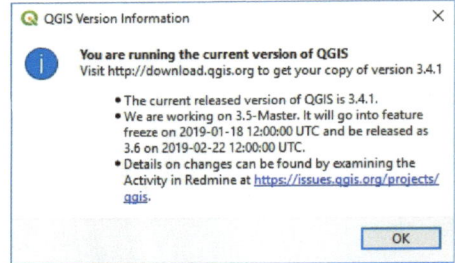

Figure 23: Checking for QGIS updates

Whenever possible you should install the "stable release" version of QGIS. Avoid installing the "latest release" version because it can sometimes be a bit flakey.

How to Make Your QGIS Interface Look Like Mine

You will find that the QGIS interface looks different from installation-to-installation. Sometimes this is because QGIS picks up the way previously QGIS installations were configured. Other times it looks different simply because the default installation changes.

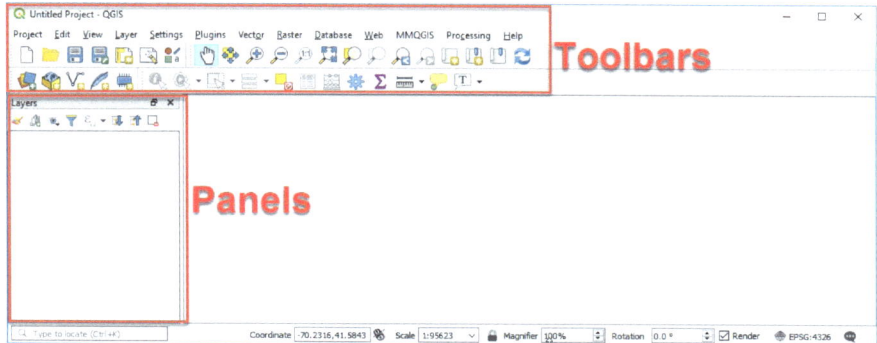

Figure 24: The format for the QGIS interface that I use throughout this tutorial.

My QGIS interface looks like this (Figure 24). Its best if you get your interface looking like mine so you can follow along with the videos more easily. To do that you need to know about QGIS Panels and Toolbars. Both are modified from within the 'View' drop-down menu…

Panel Options

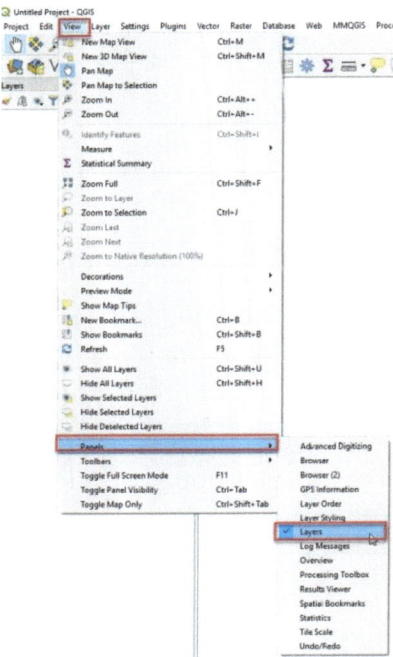

Figure 25: To make your desktop look like mine you should have only the Layers panel enabled

I only have only the Layers Panels active. For me, having other panels open takes up too much screen real estate (Figure 25)

Toolbars

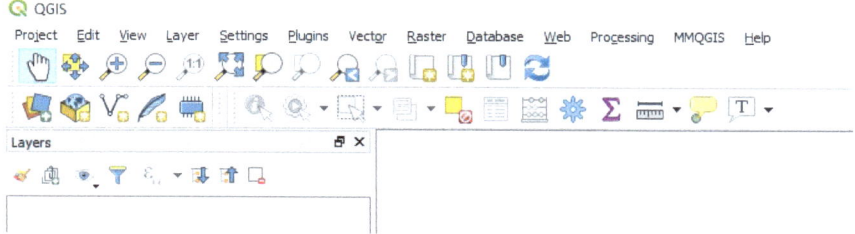

Figure 26: Here's what my toolbar looks like.

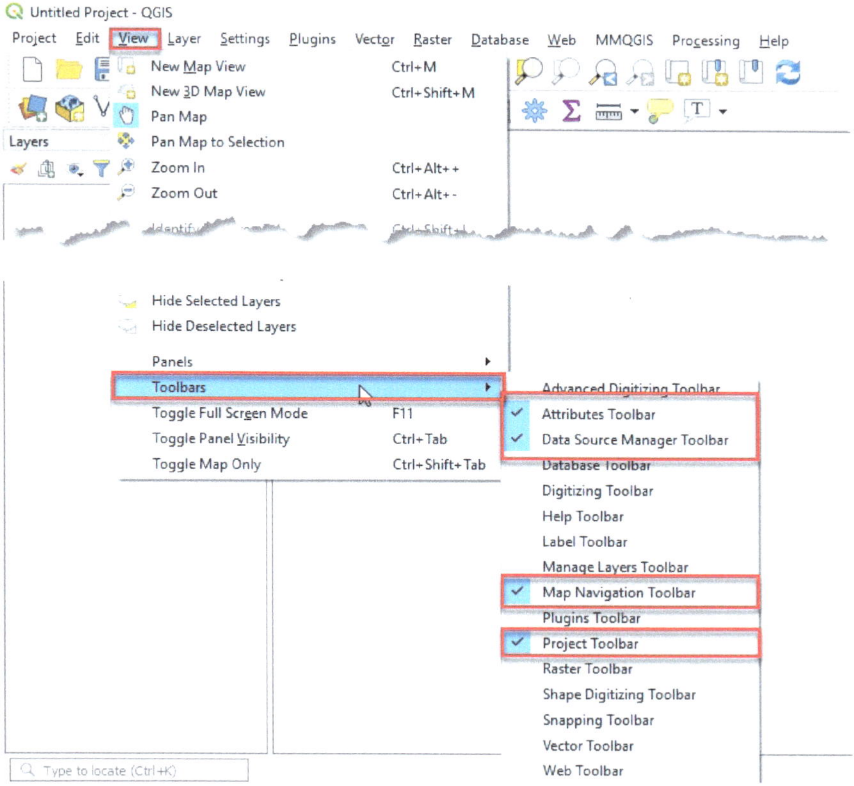

Figure 27: To make your toolbar to look like mine, enable these four toolbars.

All menu toolbars can be picked up and dragged around the screen. To make your desktop look like mine (Figure 24)...

- Enable the toolbars shown in Figure 27.

- Position the toolbars to be in the same place as mine. Holding down your left mouse button over the 7 vertical dots on the left side of the toolbar you want to move. Then drag and release it.

THE QGIS DESKTOP

An easy introduction to the QGIS desktop

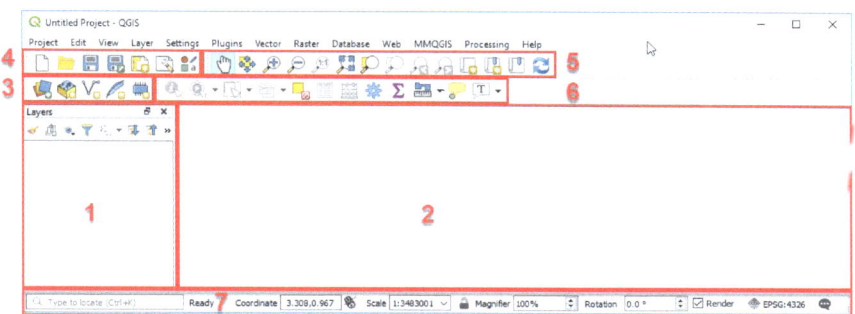

Figure 28: There are lots of components in the QGIS interface. I'm only going to talk about the ones in the red boxes in this tutorial.

In this section, I'm going to introduce you to the QGIS desktop (Figure 28). Specifically…

1. **The Map Layers Window:** This is where you change a map layer's properties (colour, etc), and the order of its display.
2. **The Map Window:** This is where maps display.
3. **The Data Source Manager:** This is where you open a GIS map.
4. **Projects Toolbar:** Projects allow you to start the day with your screen looking the same as it did at the end of the previous day.
5. **Attributes Toolbar:** How to click on something and find out things about it.
6. **Map Navigation Toolbar:** How to zoom and pan your way around a map.
7. **Bottom Status Bar:** Displays information about the map window's zoom level, cursor position, etc.

The Map Layers Window and the Map Window

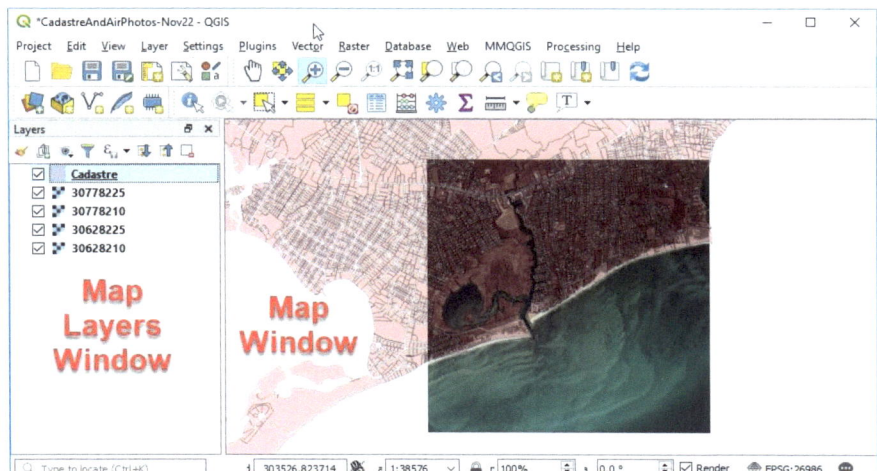

Figure 29: The purpose of the Map Window is to display maps. The Layers window is the engine room of the Map window. This is where you change the way maps display in the map window.

Now let's talk about the Map Window and Layers window.

The Layers Window

The Layers Window (1 in Figure 28) is the engine room of the Map Window. In here you can change the order that map layers display. You can also change the way a map looks on-screen.

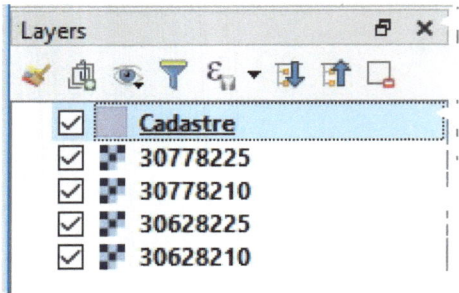

Figure 30: The Layers window.

- Click on the checkbox to display or hide a map layer.
    - You might want to print a map without one of its layers
    - You might want to display large GIS maps (ie. take lots of time for your computer to draw) only at the time you need them.
- Click on a layer and while holding down your left mouse button, you can drag and reorder it. The ability to layer maps is basic GIS functionality. It is usually best to have air photos on the bottom so you can see vector GIS maps such as roads and property boundaries on top.

Layer Properties

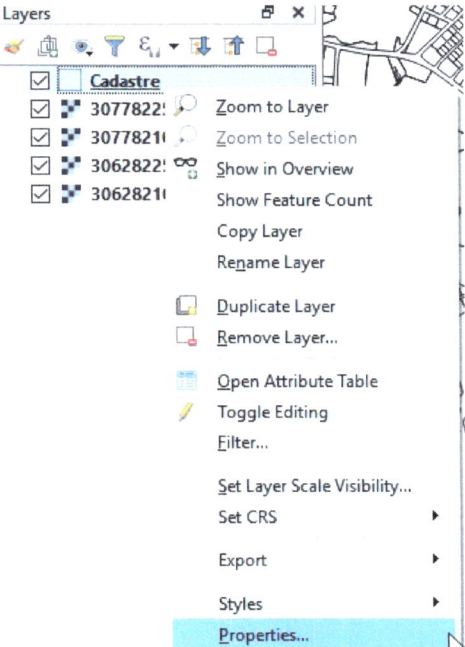

Figure 31: Right-click on a Layer in the Layers Window and a menu appears. From the Properties sub-menu, you can change the way your map looks.

Right clicking a Map in the Layers window launches the map layer properties dialog. The Symbology area is where you set the style of the layer. Cartographers spend a lot of time here.

I want you to take the time to experiment with all the options in the Style area.

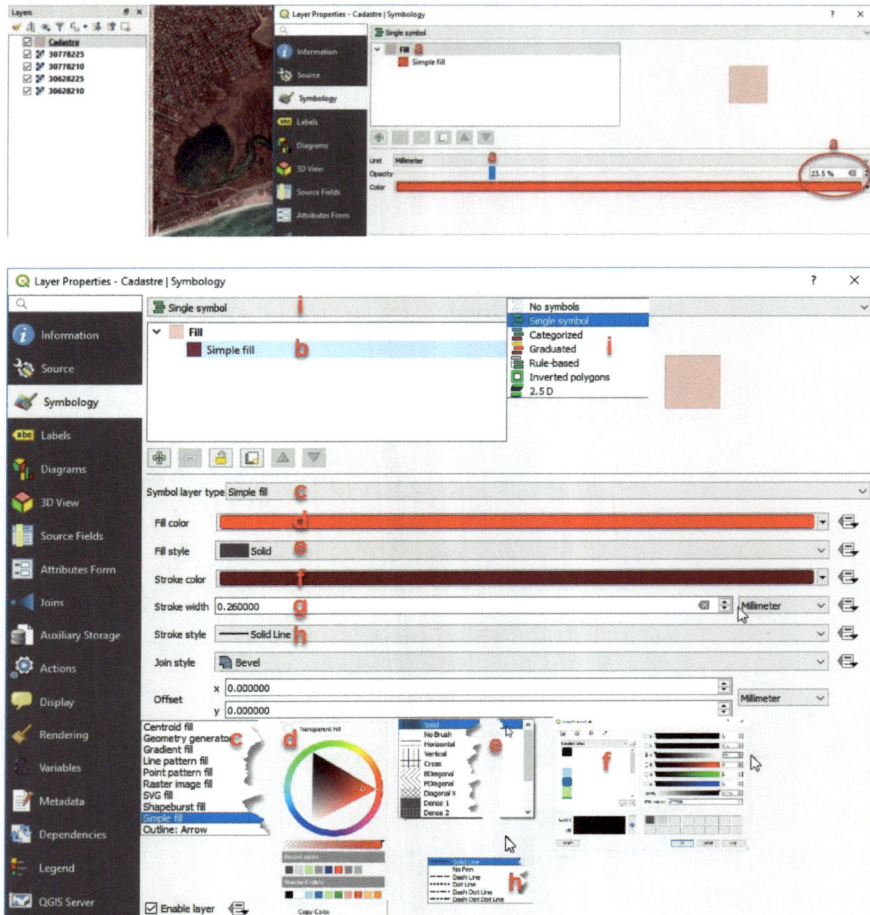

Figure 32: The Layer Properties dialog. We'll be working with the Symbology sub-menu in this lesson.

The bullet point letters that follow relate to Figure 32 on the previous page.

a) **Slider:** This allows you to adjust the transparency of the fill color. In this case you can see the cadastre (land parcels) as a transparent shade of 23.5% red. Note that I've clicked on the square labeled "Fill".

b) **Symbol Properties:** This allows you to change a whole bunch of things about the way a map layer displays.

c) **Symbol layer type:** Choose from the options shown. Stick to Simple Fill until you're confident with this dialog.

d) **Fill color:** Self-explanatory.

e) **Fill Style:** Choose from various styles (eg. cross hatching). "No Brush" makes a polygon transparent.

f) **Stroke color:** The color of a line or polygon border.

g) **Stroke width:** The width of a line or polygon border.

h) **Stroke style:** The style of a line or polygon border – dashed lines etc.

i) **Symbol:** This dropdown enables the thematic mapping feature and is the place you begin your shaded map creation journey. This is a very powerful feature. We'll be using the Single Symbol and the Categorized options in this tutorial.

The Map Window

The Map Window (2 in Figure 28) is where maps are displayed and manipulated interactively with a mouse. There is little more to say about it for now.

The Data Source Manager Toolbar

There are five buttons in this toolbar (3 in Figure 28). We'll only be talking about the Open Data Source Manager button. That's where we open vector maps and air photos.

Add a Vector Map (cadastre)

A vector map is a GIS map containing points, lines, polylines and polygons. These four object types are core to all Geographical Information Systems. I explain them in depth in the Spatial Data (Maps) section on page 7, and in the GIS Concept lessons within the Udemy video tutorial. I explain the concept of map layers in the Map Overlay section on page 12, and within the Udemy video tutorial too. All are briefly described in the glossary at the end of this book too.

QGIS can open more than 20 different types of vector GIS maps. The map we're going to open now is know in GIS-speak as a Shape file. Shape files are very simple GIS files that contain only basic shape, attribute and sometimes projection information. Object styles (eg. color and lineweight) are not recorded in this file format. This file type was first used by ESRI, the makers of the Arc range of GIS software. Over the years shape files have become a defacto standard throughout the GIS industry.

It's time to open our first GIS map layer...

Click on the Open Data Source Manager button to open the dialog.

QGIS 3 TUTORIAL FOR BEGINNERS #1: GETTING STARTED

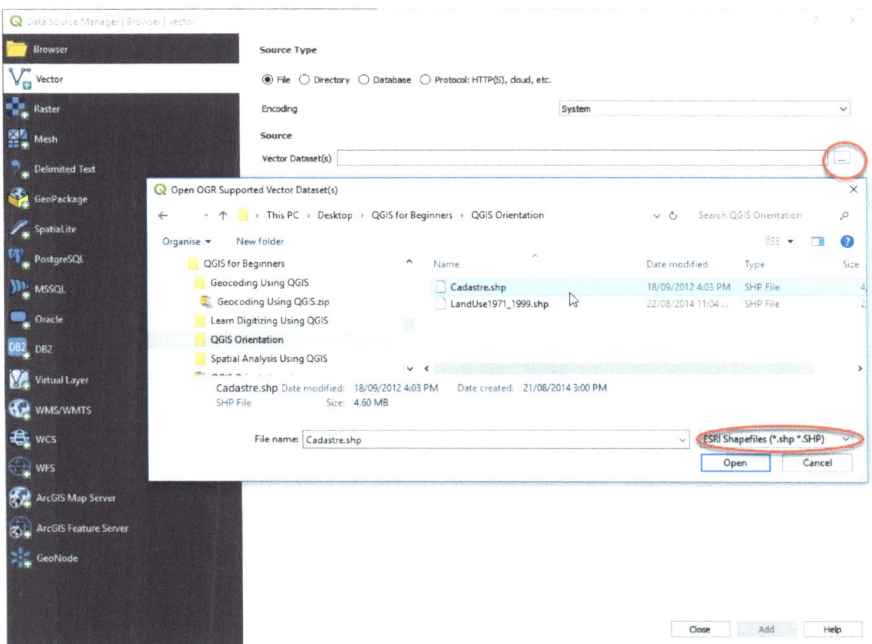

Figure 33: Click the Browse button and open the ESRI ShapeFile called Cadastre.shp.

Figure 34: Although the GIS map we're opening is a Shape file, there are many other types of GIS and data files that QGIS will open.

Add a Raster Map (air photo)

A raster map (for the purpose of this tutorial) is an air photo or satellite image. Air photos are in common use these days – you see them in web mapping systems such as google maps and bing maps. Air photos are bread-and-butter for local planning agencies who use them mostly for visual analysis. The important thing for us to understand is that for a photo or satellite image to be useful in a GIS, it must be geo-referenced so that it can be related to other maps in the GIS.

So, lets open our first GIS raster map layer…

 Click on the Open Data Source Manager button to open the dialog.

QGIS 3 TUTORIAL FOR BEGINNERS #1: GETTING STARTED

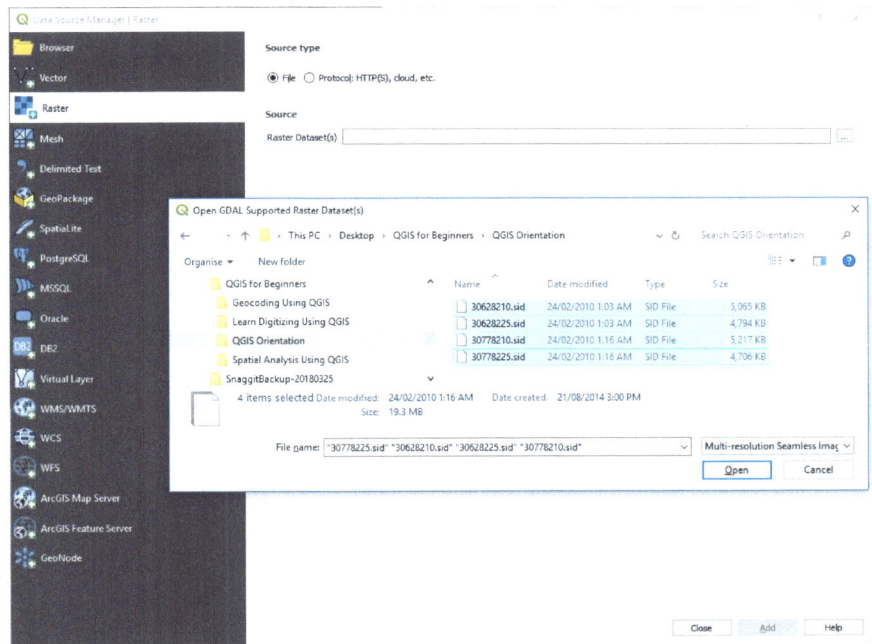

Figure 35: Open the four files with the "sid" extension. These are MrSid file, a common raster format used in GISs. You can open them individually or all four at once. To select more than one file, experiment with depressing either the Shift or Ctrl buttons on your keyboard while clicking on the files.

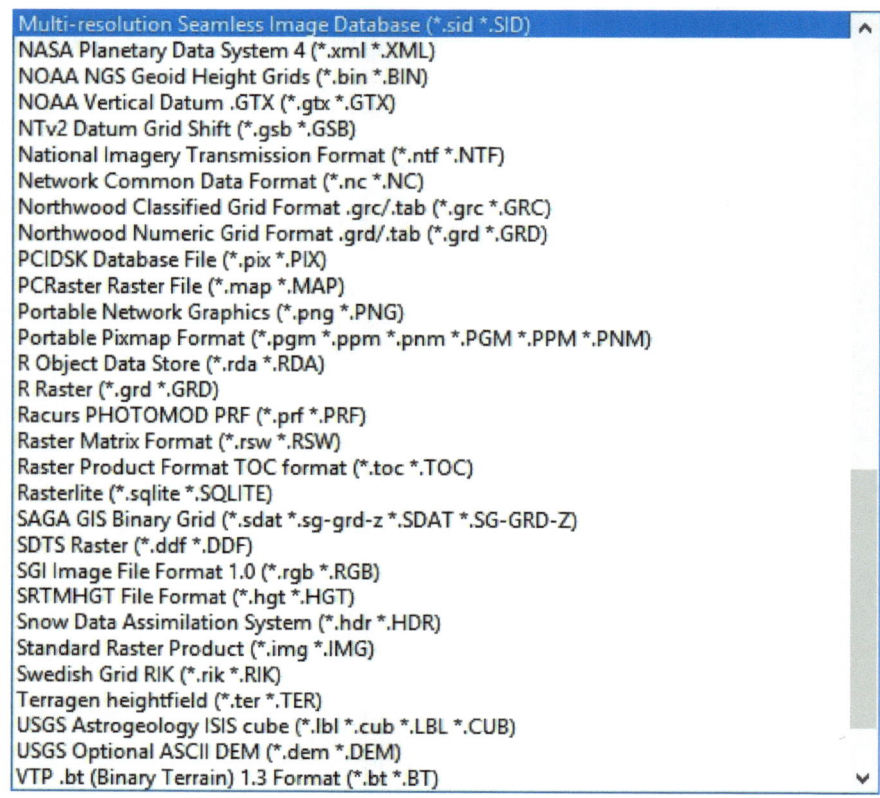

Figure 36: Although we just opened a MrSid file, QGIS can open tens of graphic formats.

QGIS 3 TUTORIAL FOR BEGINNERS #1: GETTING STARTED

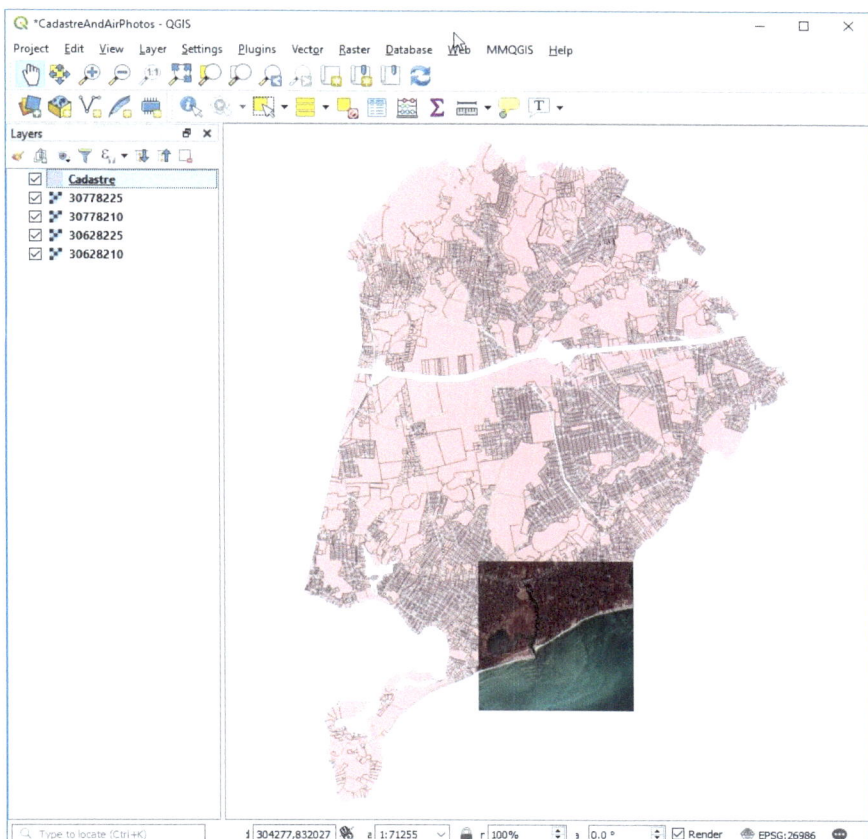

Figure 37: The four air photos open up. If they're on top of the vector map then click on the cadastre map, hold down your left mouse button, and drag it to be on top of the air photos.

The Projects Toolbar

Projects (4 in Figure 28) are a useful feature of QGIS. It's a feature that most, if not all, geographical systems have. Put simply, when you save a project, you're saving your present work environment. This means that you can save a project before you shut down your computer at the end of the day. At the start of the next day you can open the saved project and pick up where you left off. The buttons relevant to this tutorial are…

	New Project	Create a new project. Each project has its own coordinate system, and various settings such as zoom levels, Print composers, text, scalebar, north arrow and copyright styles, plugins, etc.
	Open Project	Open an existing project and pick up where you last left off.
	Save Project	Save the project you're working on. This is where you give a project its name.
	Save Project As	Save a project as a new name. This is useful when you need to experiment with variations of styles, map zooms, and so forth.
		Imagine you named your Project with a version suffix such as Project-v1 thru Project-v7. If you discovered an error had been introduced in v6, then at least your starting point could be v5 rather than the very beginning!

QGIS 3 TUTORIAL FOR BEGINNERS #1: GETTING STARTED

The Map Navigation Toolbar

The tools in the map navigation toolbar (5 in Figure 28) allow you to use your mouse to move a map around the screen. The buttons relevant to this tutorial are explained below…

 Pan Map — When you've got a map on the screen, choose this tool, hold down the left mouse button and then drag the map around the screen.

 Pan Map to Selection — Select a geographic object and click this button to move it to the centre of the screen.

 Zoom In — Click on the map and while holding down the left mouse button, drag it to create a square. Release the mouse button and QGIS will zoom into that area.

 Zoom Out — Click on the map with the left mouse button and the map will zoom out.

 Zoom to Native Pixel resolution — I have to be honest. I have never used this. A search around the net indicates its used to zoom to an optimum scale for a raster file.

 Zoom Full — Zooms to the full extent of the map you have highlighted in the Layers window.

 Zoom to Selection — Zooms to whatever it is that you have selected. This is a really powerful feature because it allows you to find geographical objects you've selected via database query. Don't worry about this too much for now. It will make more sense later on in the tutorial.

 Zoom to Layer — Zoom to any layer you have selected in the Map Layers window.

Zoom Last — Go back one zoom level.

	Zoom Next	Go forward one zoom level.
	Refresh	If a map fails to draw properly this button allows you to refresh your screen.

The Attributes Toolbar

Figure 38: The Attributes Toolbar.

Let's talk about the Attributes Toolbar (6 in Figure 28). We'll address…

  i. Finding information about an object on your map.
  ii. Different ways of selecting items on your map.
  iii. Selecting some, or all of the features in your map.
  iv. Unselecting a Selection.
  v. Tables…
   a. Different ways of selecting items in the table lying behind your map.
   b. How to modify the table that lies behind your map.
   c. An introduction to the link between maps and tables.
  vi. Measuring distances and areas interactively.

Identify Features Button

Identify features   Identify features button (i in Figure 38) allows you to click on a map feature and find out information about it

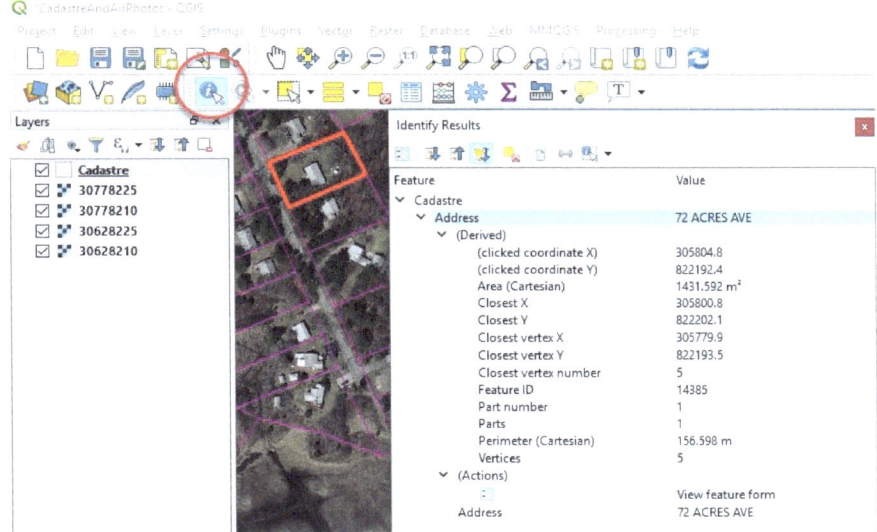

Figure 39: When you click on a parcel you get to find out its address as well as how big it is.

Now we're really starting to get into the nuts and bolts of what sets GIS apart from maps in graphics systems. In a GIS there's a database that lies behind each map, and that database has two types of data in it...

1. **Map (spatial) data:** You can select map features such as a land parcel and discover geographical information about it. For example, how big is it and what's its perimeter?

2. **Attribute data:** This is information about the item you've chosen. For example, it might be owned by Mr and Mrs Smith, or be on Smith Street.

The Geographical Selection Button

The Geographical Selection Button (ii in Figure 38) allows you to select geographical objects in five ways.

Selections are important because the geography you select gets isolated from the other objects in your GIS database. That means that you can to do extra things to only them, without affecting other geographical objects or other rows in your table. You could update the attributes of individual rows in your table, the colors of objects in your map, or create a report based on your selection. For example, a local government employee could use their mouse to select all the properties in a street frontage and then notify the owners of planned roadworks.

To activate these buttons you first need to click on the Cadastre map in the Map Layers window and highlight it. When you've done that, the down arrow to the right of the Selection button reveals all the Selection possibilities (Figure 40). Click on one of the selection buttons and then on a map feature. I explain the way each operates below.

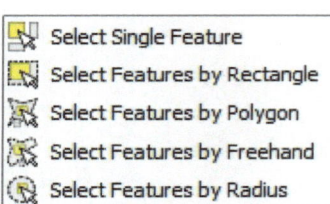

Figure 40: All the Geographical Selection possibilities.

	Select single feature	a	Left mouse click to select a feature. When you click the down arrow, a drop-down menu appears (Figure 40).
	Select within rectangle	a	Depress the left mouse button and drag to create a rectangle. When you release the mouse button everything within the rectangle will be selected.
	Select within polygon	a	Click-click-click with the left mouse button until the polygon you've drawn surrounds the objects you want to select. Press the right mouse button everything within the shape will be selected.
	Select within freehand polygon	a	Hold the left mouse button down and drag the mouse to draw a shape. When you take your finger off the mouse button everything within the shape will be selected.
	Select within radius	a	Hold the left mouse button down and drag until the circle you've created surrounds everything you want to select. When you take your finger off the mouse button everything within the circle will be selected.
	Unselect		Click on this and everything you've selected will be unselected (iv in Figure 38).

Selecting Some or All Features in Your Map

The Selecting Some or All Features in Your Map option (iii in Figure 38) allows you to make selections using a database approach. The drop-down menu to the right of the button reveals all four selection possibilities.

Figure 41: All the Database Selection possibilities.

Highlight the cadastre map in the Layers area first. That tells QGIS that we want to select objects from the cadastre map.

Select features by value

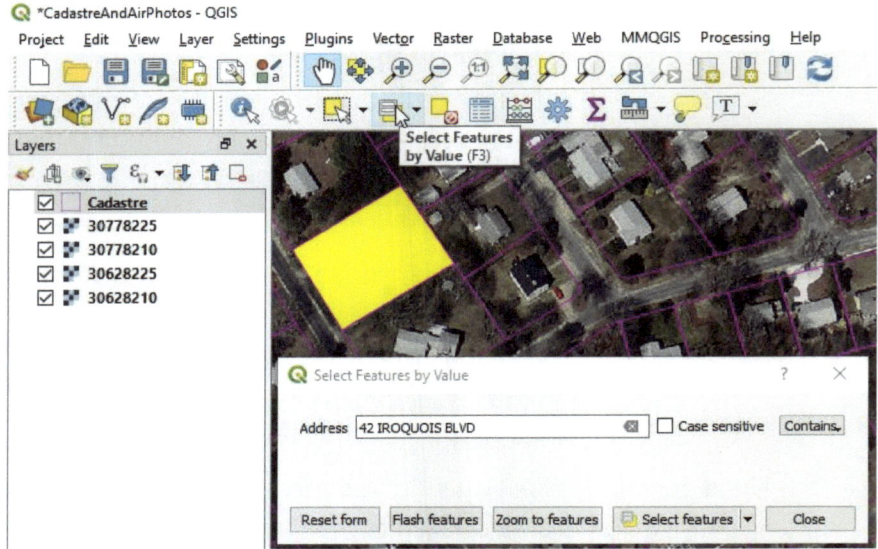

Figure 42: The select features by value form.

Choose to Select Features by Value and a form appears. Type in an address and choose to either zoom to the feature or select it (Figure 42).

Select Features by Expression

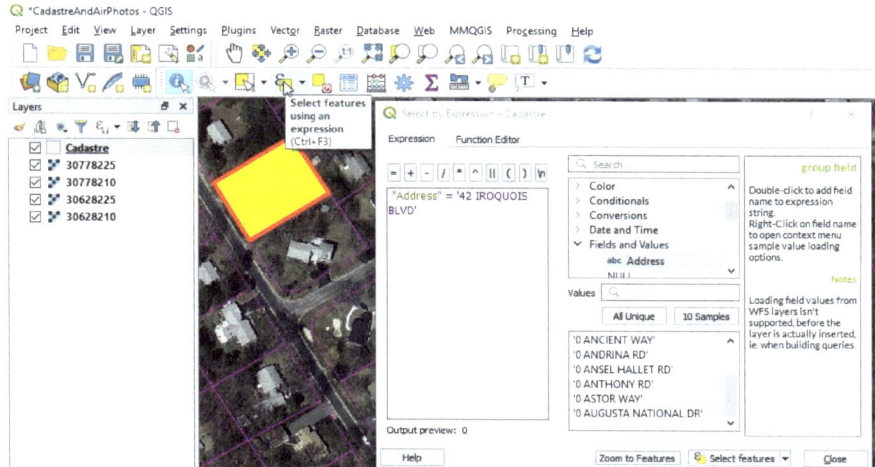

Figure 43: The Select Features by Expression button launches an SQL dialog.

The Select Features by Expression button launches an SQL dialog (Figure 43). This is a sophisticated way of selecting a geographical object. A detailed discussion is beyond the scope of this beginners tutorial.

The Invert Selection button

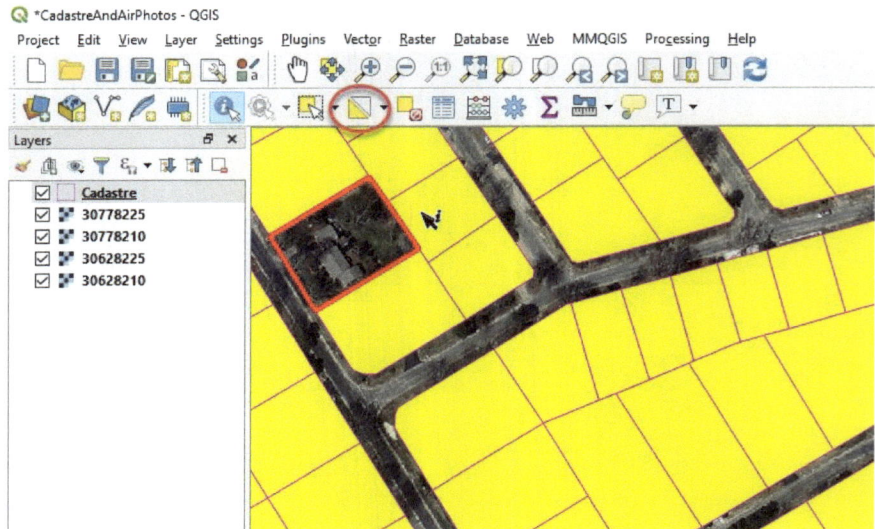

Figure 44: Invert Selection button.

This button inverts your current selection (Figure 44). In this case it has selected every cadastral parcel excepting the one we had selected earlier.

The Select All Features button

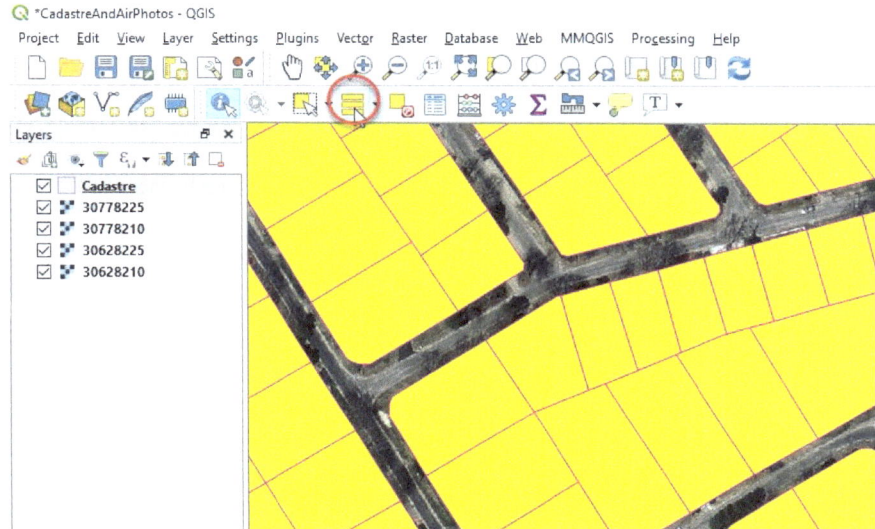

Figure 45: The Select all Features Button.

As the name suggests, this button allows you to select everything in your map (Figure 45).

Open Attribute Table Button

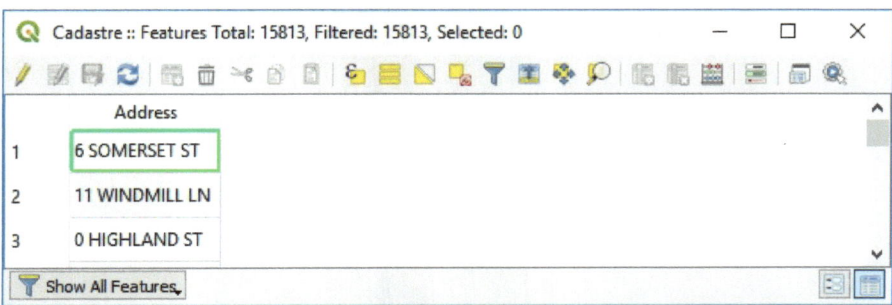

Figure 46: There are 15,813 parcels of land in the Cadastre map and therefore 15,813 rows in its attribute table.

The Open Attribute Table Button (v in Figure 38) is where QGIS really starts to get interesting! It is enabled once you've chosen a map layer from within the Map Layer window. With this option a whole bunch of GIS power comes your way. For you to understand this, I'm going to spend some time talking about tables. I'll do this in very general terms.

Every GIS map has an attribute table attached to it. The attribute table contains information about each object in the map. In the case of our Cadastre map, the table contains Address information. Every one of the 15,813 rows in the Cadastre table contains the address of one parcel of land. In a GIS there is also geography attached to each row.

When you've got geographical information stored this way you can do all sorts of things with it. An obvious one with our Cadastre table is to search for a land parcel at an address – much like you would search for an address in google maps. OK you're thinking, so if it's just like google maps then why not use google maps? Well, there's three parts to this answer...

1. google maps mostly has display functionality in it, and
2. very little analytical functionality, and
3. only for maps of Google's choosing.

In contrast, QGIS gives you powerful analytical functionality for your *own* maps - maps such as census maps, vegetation maps, or any other sort of map you would like to add or create yourself.

How to Open an Attribute Table

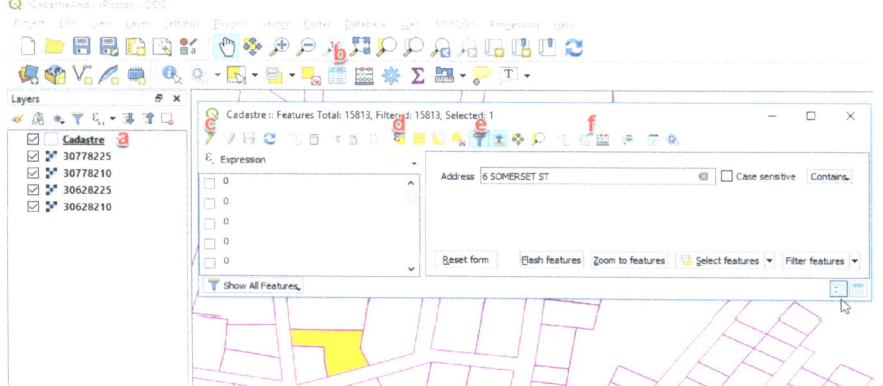

Figure 47: With a vector map highlighted in the Map Layer Window (a) the Open Attribute Table button is activated (b). From within the resulting dialog box you can view and manipulate the database that lies behind the map you highlighted.

Let's now open the cadastre table and look at some of the functionality available in the Attribute Table dialog box…

	Open Attribute Table *(b)*	Click on the Cadastre Map layer *(a)* to activate the Open Attribute table button *(b)*. Click on the button and the Cadastre attribute table is displayed.
	Toggle editing mode *(c)*	This button enables and disables the editing mode.
	Select features using an expression *(d)*	This button launches a dialog that allows you to make a Selection based on search criteria. This is an advanced concept dealt with in more detail in my "GIS for Beginners #4: Learn Geocoding in GGIS 3" tutorial.

	Select features using a form *(e)*	Similar to the Select Feature by Expression button in Figure 42.
	Open field calculator *(f)*	Similar to the Select by Expression button in Figure 43. If editing mode is enabled *(c)* you can create new fields / columns here

How to Search

Figure 47 and Figure 48 show the results of a search for a partial address "6 SOM" that I entered into the address area. After clicking the Search button QGIS selected "6 SOMERSET ST".

Things you can do with Selections.

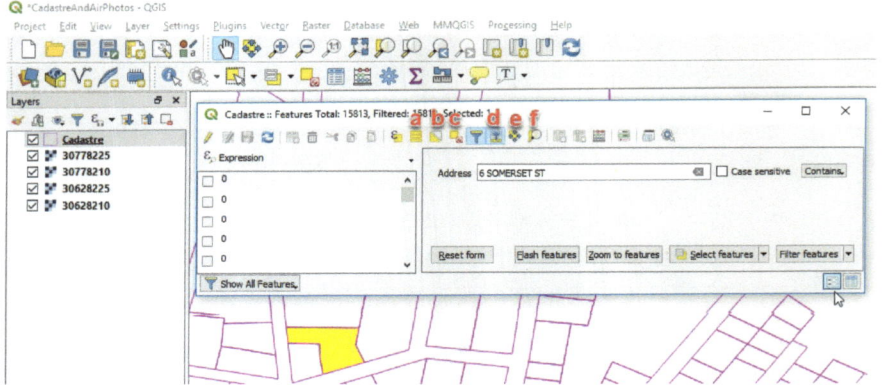

Figure 48: You can do all sorts of things with a Selection, and there's a button for each thing.

The search we did in Figure 47 and Figure 48 yielded "6 SOMERSET ST" so this search is said to have been successful. The parcel at "6 SOMERSET ST" in the cadastre map is now the *Selection*. You can do all sorts of things with a Selection, and there's a button for each thing…

	Select all *(a)*	Selects all the rows in the table.
	Invert selection *(b)*	This option unselects your selection and then selects everything else. So, if you were to click this button "6 SOMERSET ST" would be Unselected and the other 15,812

	Unselect All *(c)*	rows in the table would be Selected. Unselects any rows you've selected. Clicking on this button will unselect the row with the "6 SOMERSET ST" address.
	Move Selection to top *(d)*	When the Cadastre table is sorted alphabetically, "6 SOMERSET ST" is on row 12,694. Clicking this button moves our selection to the top of the display so you can see it. If a different search returned three addresses – one from the beginning of the table, another from the middle and another from the end, clicking this button would make them appear as rows 1, 2 and 3.
	Pan to Selected rows *(e)*	This is database selection and geographical selection working together. Clicking this button keeps the current zoom on the map, and *pans* the selection.
	Zoom to Selected rows *(f)*	Another example of database selection and geographical selection working together. Clicking this button *zooms* the map into "6 SOMERSET ST".

You can modify the table too

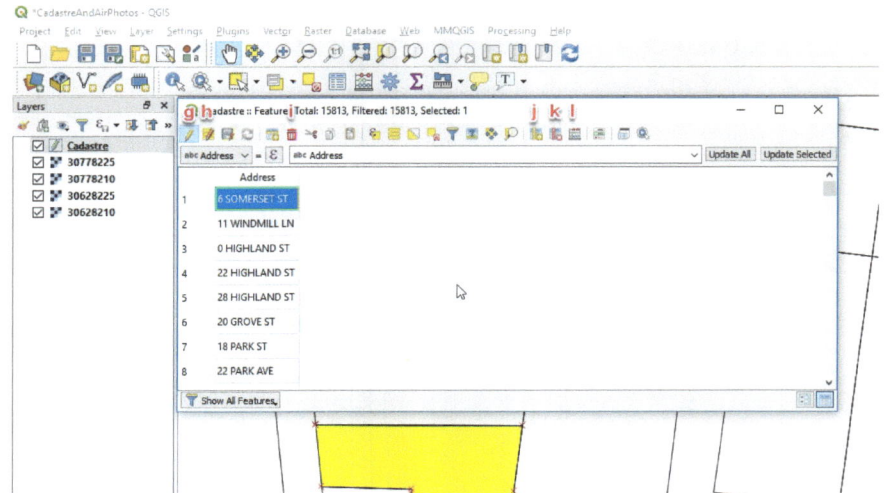

Figure 49: *The modify table buttons.*

These buttons allow you to add and delete columns, make sophisticated selections based on either the values in the columns and/or geographical characteristics of the map (eg. land parcels >1ha), delete the rows you select, and save any changes you might make to the table. Viewed in isolation they don't sound very inspiring, but as a part of a GIS toolkit they are very powerful!

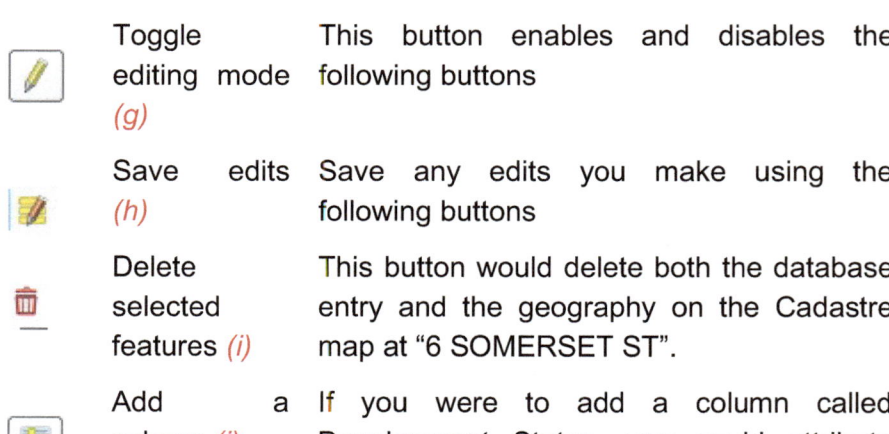

	Toggle editing mode *(g)*	This button enables and disables the following buttons
	Save edits *(h)*	Save any edits you make using the following buttons
	Delete selected features *(i)*	This button would delete both the database entry and the geography on the Cadastre map at "6 SOMERSET ST".
	Add a column *(j)*	If you were to add a column called Development Status, you could attribute each parcel with a value of either "developed" or "vacant".

	Delete a column *(k)*	Self explanatory
	Field Calculator *(l)*	Having added and populated a Development Status column the Calculator is where you'd ask QGIS how many blocks are vacant and what's the total area of the vacant lots.

Measurement Tools

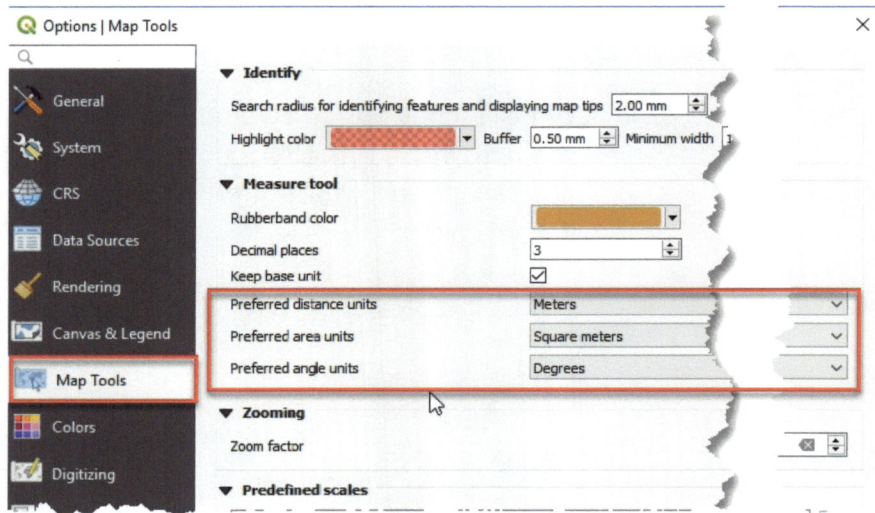

Figure 50: You change the measurement units in the Settings -> Options -> Map Tools dialog box.

The following three tools (vi in Figure 38) can be used to measure things on your computer screen. For example, the distance between two things or the area of a building on an air photo. You can set the measurement units to be either feet or metres via the Settings -> Options -> Map Tools menu path (Figure 50). The example above shows metric units, so lengths are shown in metres and areas in square kilometres.

All the measurement tools are all "dynamic" meaning that the measurements in the dialog box respond to you moving the mouse around.

The three measurement options are accessed from the down-arrow to the side of the measurement button.

QGIS 3 TUTORIAL FOR BEGINNERS #1: GETTING STARTED

 Measure line — Measure the length of a line you draw with your mouse.

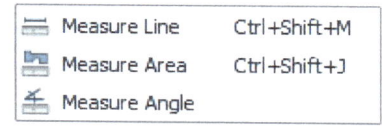

Figure 51: All the measurement possibilities

 Measure area — Measure the area of a shape you create with your mouse.

 Measure angle — Measure the angle of one line from another

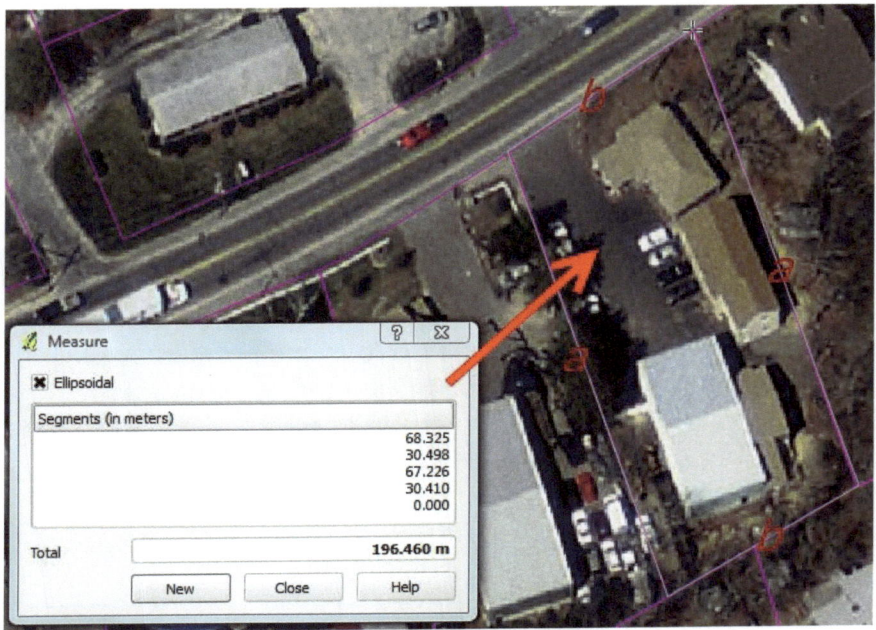

Figure 52: This is an example of the line measurement tool. The length of each line segment and the total length of all the line segments are shown in the dialog box that pops up when you begin measuring. Even though this block is not square, the sides labeled "a" and the sides labeled "b" are similar lengths to each other. All the measurement tools are "dynamic" meaning that the measurements in the dialog box respond to you moving the mouse around.

To measure the length of a line (Measure Line): Click with the left mouse button to start measuring. Move your mouse and click with the left mouse button again to tell QGIS the point you want to measure – you've just created a measurement line "segment". QGIS will display the distance between the two mouse clicks. Another click on the left mouse button will create another measurement segment again. A single right click stops the measuring. Two right clicks zeros the measurement.

To measure the area of something (Measure Area): Click with the left mouse button to start measuring and click with the left mouse button again and again until you've crawn the area you want to measure. A single right click stops the measuring. Two right clicks zeros the measurement.

To measure an angle from a line (Measure Angle): Click with the left mouse button to start a line and click with the left mouse button again to create a line from which to measure an angle. Another click on the left mouse button will open a dialog box within which the angle measurement is shown. A single right click stops the measuring. Two right clicks zeros the measurement.

Bottom Status Bar

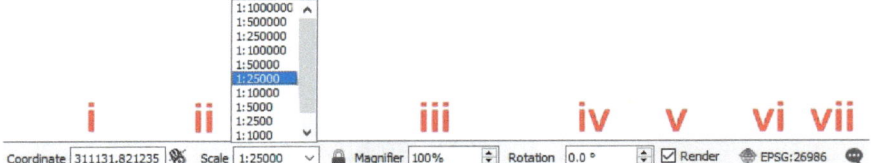

Figure 53: The bottom status bar is host to a bunch of information about the Map Window.

The bottom status bar (7 in Figure 28) is host to a bunch of information about the current map window. This includes…

i. The map XY coordinates of the mouse arrow

ii. The scale of the map display. You can choose from one of 11 predefined scales. Otherwise have a custom scale by just zooming however you want. You can define your own scale in the Settings -> Options -> Map Tools dialog box.

iii. Another zooming tool

iv. Rotate your map plus or minus 360 degrees. Useful if comparing to uncorrected air photos.

v. The ability to render (redraw) your map, or to stop the redraw if its taking too long.

vi. The current coordinate system. In this case EPSC:26986 represents the NAD83 / Massachusetts Mainland coordinate system

vii. System messages.

Visual Interpretation of the Vector Property Map and Air Photo Map

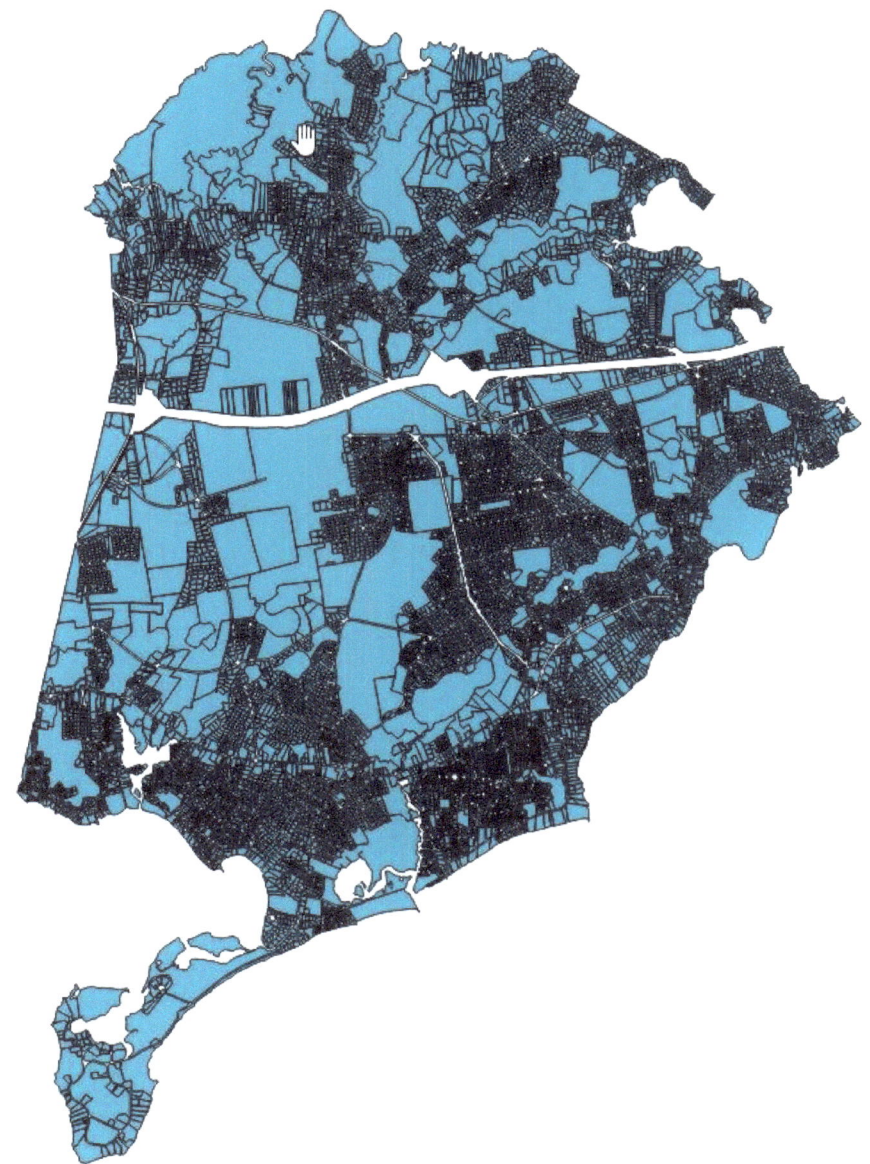

Figure 54: Finally, a map! Just looking at this map I can imagine numerous problems that GIS could help with - environmental, agricultural and planning uses come to mind.

Those who've seen me around the web (especially Udemy, Quora, YouTube and Facebook) will be familiar with my belief that GIS is far more than just a technology. The best GIS projects are done by people who have an understanding of both GIS and the problem it is being applied to. In other words, they use GIS as a tool!

I would like you to start thinking of yourself as a GIS analyst. Don't just take maps and air photos at face value. You need to put your thinking caps on, look for "patterns" in maps, and then think about what those patterns might mean. With that in mind, let's now use the tools we've learnt to delve further into the shapes, patterns and colors in the air photography and cadastre. There are many tricks for interpreting both vector maps and air photos. Most are logical when they're pointed out to you.

I want you to look a little more closely at the cadastral/property map (Figure 54 or on-screen) and imagine some of the issues that might be important in the area. Here's four that I've come up with…

- **Urban areas:** There are many small blocks and most times you would be right in thinking this would be a suburban area. There also appear to be areas of slightly larger blocks suggesting rural-residential uses. There will be plenty of urban planning issues. For example building regulations, social service provision, planning for schools and road maintenance.

- **Non-urban areas:** Some of the large blocks in the middle might be used for farming, forestry, or public open space. Undoubtedly there will be planning pressures on such large parcels of land in close proximity to urban areas.

- **Inland waters:** Many blocks have rounded edges. This suggests the presence of water bodies. There will also be water quality management issues.

- **Coastal areas:** The bottom of the map (south) is almost certainly a bay, so there are likely to be coastal wildlife and urban water runoff issues.

Often maps have a story to tell if you take the time to look closely and think about them. The stories I have just touched on would need to be confirmed in the field, and potentially aided by census, planning, infrastructure and environmental maps.

Figure 55: Air photo interpretation.

Let's now look at the air photos (Figure 55 or on-screen) to see if there might be stories they could tell. I like to think in terms of "surrogates". All would need to be confirmed by fieldwork…

1. Pond area:
    - The large blocks are large because they're poor land and probably subject to inundation. The drains provide a clue that the area is swampy. Notice that the drains are

straight and close together. These are not natural. However, the land must be useful in some way, otherwise it would not be worth the effort to drain it!

2. Inlet area:

   - Alongside the inlet there are permanent boat moorings, so demographically it is likely to be an affluent area.

   - The water in the inlet must be deep. Otherwise there'd not be a mooring so far up.

   - Green lawns mean that gardens are being maintained and suggests greater water usage in this area. In times of drought water utilities might target these areas for watering restrictions. There might also be issues relating to nutrient runoff into the inlet. Nutrients can cause algal blooms and consequent fish kills. Algal blooms have been known to kill livestock that drink it.

3. **New estate:** You can tell this because...

   - Lawns are green.

   - Roads are new.

   - The air photo was taken in summer:
     - In most places the lawns are brown.
     - Trees seem to be suffering so maybe there are drought issues. Its also possible that that's just how the trees in the area look in an air photo!
     - Mostly single storey homes. I can tell this because they cast small shadows.
     - Lack of swimming pools may be cultural, but also socioeconomic.

- General:
    - The condition of road infrastructure often provides a clue to the age of a development. When you zoom in, there are some areas where the roads are very cracked and have many repairs.
- Soils:
    - Sandy in places due to dry lawns.
    - Clays in other places due to lakes.

SHADING A MAP

How to Shade a Map in QGIS 3

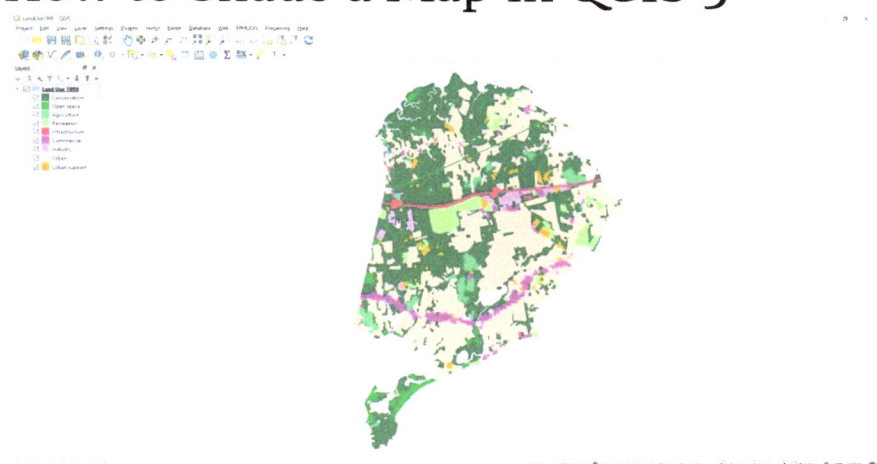

Figure 56: This is what the finished map looks like.

In this section we're going to shade a map using land use category data. The area is Yarmouth Massachusetts. You can open the LandUse1999 project and see for yourself if you like. Assuming that you've installed the tutorial dataset, it is in the "..\QGIS for Beginners\QGIS Orientation\" folder on your Desktop.

This section of the tutorial has a fair bit of breadth to it. I'm going to talk about…

- How each column in a table can be used to create to a different shaded map.
- How to generalize data categories. And why sometimes you need to do this.
- How to shade a map using data categories.
- How to choose a colour scheme for your map.
- How to use other maps to validate the Land Use map.
- How to map Land Use Change through Time (time series mapping).

Background: GIS Tables

In this section, I'm going to talk about the relationship between the data in the table attached to your GIS map, and the thematic map that we'll be creating.

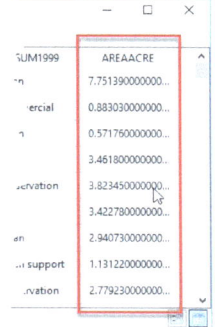

Every GIS map has at least one column of data in its accompanying table. Often GISs automatically produce columns of data such as the AREAACRE column in (Figure 57).

Every column of information in a GIS table can become a thematic (shaded) map. And every row in a GIS table relates to an item of geography in your GIS map. That means that each item of geography can be shaded according to the value of a data cell.

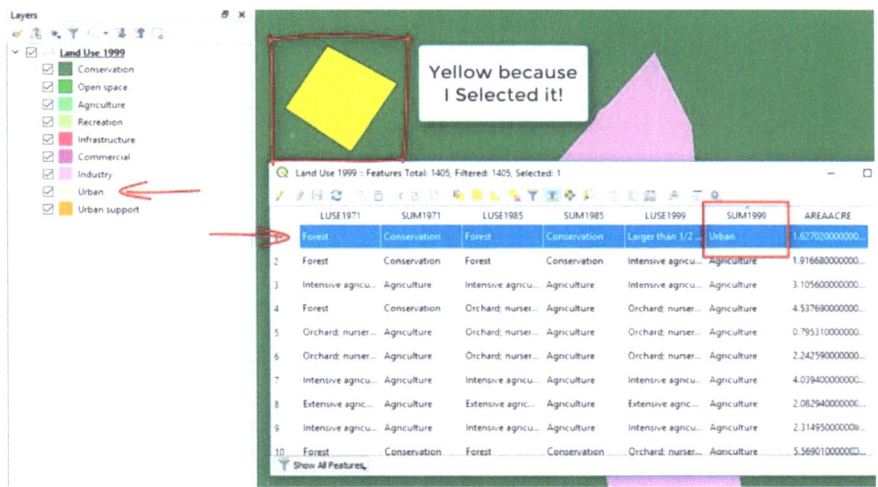

Figure 58: The relationship between tables and maps.

Had the polygon in Figure 58 not been selected (its yellow now), it would be the same colour as Urban in the legend (pink). In the table that lies behind the map, this *urban* polygon is represented by a row. The theme being mapped comes from the SUM1999 column. The value in the cell at the column/row intersection is URBAN.

The shaded thematic map in Figure 56 was created from the SUM1999 column. That column contains category data. Every column in this table could be a shaded thematic map. The AREAACRE column is numeric. The thematic mapping of numeric data requires a very different approach than mapping categorical data (this tutorial).

Later in this tutorial I'll show you how to map the Land Use in 1971, Land Use in 1985, and Land Use in 1999 columns. But first I want to talk about the need to generalize the data we'll be mapping.

Background: Map Generalization

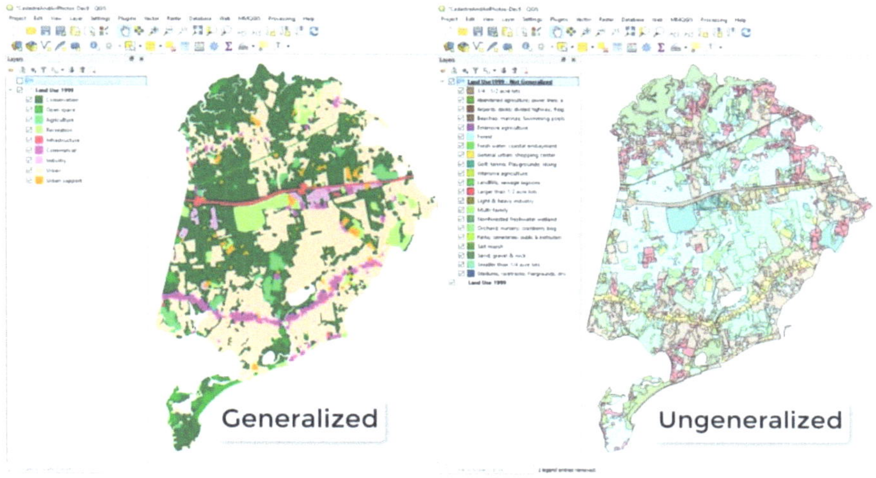

Figure 59: The Generalized map compared to the un-generalized map.

Data generalization is a very important GIS skill. Generalized maps are often simpler, easier to understand and more communicative. Data generalization does not have to result in the loss of detailed data. I don't show you how to map numeric data in this tutorial. I do talk about it some more in a couple of pages time.

The appearance of our Land Use map is heavily influenced by the way the data in it are categorized. There were two things that I considered before I generalized the data...

1. I wanted to simplify this tutorial and minimize the length of the Udemy videos.
2. Because the Land Use map is a time series map, I wanted the categories to be consistent in all three Land Use years.

I generalized the 21 data categories in this GIS map to be 9 categories. Let's have a quick look at the before-and-after versions of the maps (Figure 59)...

- **The un-generalized map has 21 categories:** The categories have been shaded according to a QGIS default colour scheme.

- **The generalized map has 9 categories:** The categories have been shaded based loosely on a planning department colour scheme.

Why would you want to generalize a very detailed map such as this is? Its about audience!

- **Researchers:** A map with detailed categories is likely to be very important to them.
- **Decision makers:** In a meeting situation, a map with detailed categories is likely to cause confusion. The person presenting the map is in danger of losing their audience.

The logic you use for generalizing maps will vary from project-to-project. My logic for generalizing this Land Use map was that I felt there were too many categories describing similar things. So I combined categories in the following ways…

- **Urban:** "Smaller than 1/4 acre lots", "1/4 - 1/2 acre lots", "Larger than 1/2 acre lots".
- **Agriculture:** Intensive agriculture, Orchard; Nursery; Cranberry bog.
- **Conservation:** Forest, Fresh water; coastal embayment, Non-forested freshwater wetland, Salt marsh.
- etc.

Background: Time Series Mapping

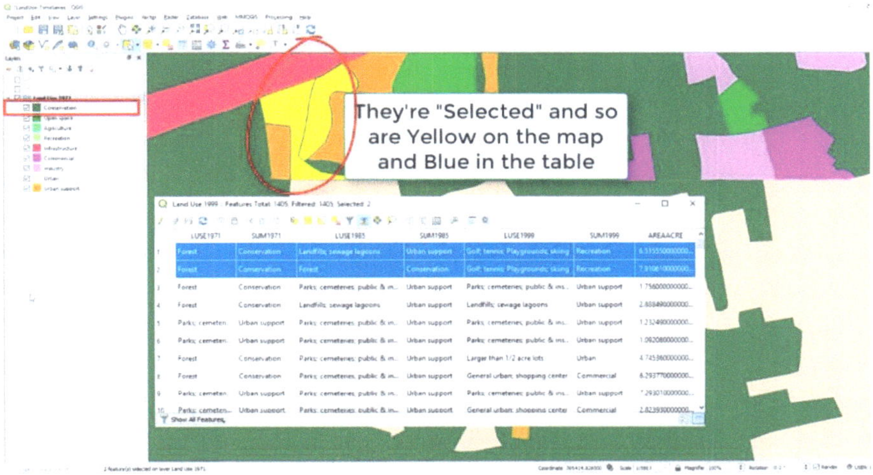

Figure 60: Time series mapping is an important skill.

Time series mapping is an important GIS skill. It allows us to show change, and can give context. Any single year of our Land Use map viewed in isolation is not as meaningful as it is when its compared to maps from other years.

Instead of having three Land Use maps, one each for 1971, 1985 and 1999, I have combined the three maps into a single map. Each year is represented by a column of data.

I pre-processed this map using Map Overlay techniques. Such techniques are beyond the scope of this tutorial, but I do discuss them in my Spatial Analysis Tutorial (there's a discount coupon for this at the end of this book).

Here's a worked example of what I mean by pre-processing. Look at the two yellow Land Use polygons in Figure 60…

- Both were forests in 1971.
- In 1985 one stayed forest but the other became some sort of urban support land use (sewerage or landfill). The Urban Support portion of the polygon needed to become a separate polygon.

- In 1999, both polygon's land use changed to be recreation, so, rather than GIS polygons requiring change, both were relabelled in the data table.

How to Shade a Map using Categorized Data

Let's shade the map. Open the LandUse1971_1999 map if you haven't already. In the layers window, right-click and go to Properties. Click the Symbology tab. This dropdown-list is really important.

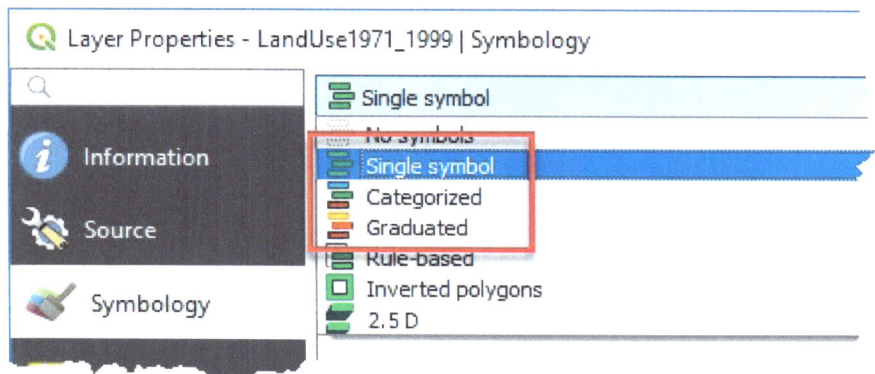

Figure 61: The three options for shading a map - single symbol, category and graduated.

I'll start by giving you a brief overview of the three options in the dropdown list (Figure 61). Then I'll show you how to shade the land use map using the Categorized option...

Single Symbol: Use this option to shade your map all the same colour. I use this for doing simple map overlay. For example, I might overlay property outlines onto an airphoto.

Graduated: Of the three shading options, this is by far the most tricky. It's a shading technique that will only work on columns that contain numerical data. So for example, a graduated map of...

- Census data might represent how many people live in each suburb.
- For an orchard, tons of fruit per acre.
- For wildlife, density of critters per acre.

Too often, inexperienced operators create Graduated Maps without understanding what they're doing. They create default GIS-assisted maps on the basis of how they look, and fail to understand that their map needs to tell a story. Sometimes they get it right by accident. Most times their maps are either misleading or wrong.

There are a whole bunch of issues that go with mapping numerical data. Things like…

- **What colour scheme should you use?** I often use a traffic light inspired scheme where red = bad, amber = OK and green = good.
- **How many categories should there be?** Too few makes a map too generalized. Too many make a map too detailed.
- **Should the classes be equal widths?** If they are, some classes may not have any data in them.
- **Should there be the same number of GIS objects in each category?** This can make the width of the classes seem illogical.
- **Should the categories reflect the breadth of the dataset?** Or should the categories focus on a small part of the dataset?
- **Time series:** And how do you go about choosing a classification system for maps that need to be compared over time?

There are so many issues associated with creating graduated maps. A detailed discussion involves the use of graphing as well as all the considerations I just listed. It is a topic for a tutorial on its own. If enough people request such a tutorial I'll happily create one.

Categorized: This is what we're going to do today. This is, as the name suggests, shading a map according to data categories (or names). Our categorized data are Land Use categories. If you were doing a qualitative study such as Vegetation Quality or Land Suitability, the categories might be something like High, Medium and Low. In other words…

- High, Medium, or Low vegetation quality, or
- Land that is Highly Suitable, Moderately Suitable or Unsuitable for a crop.

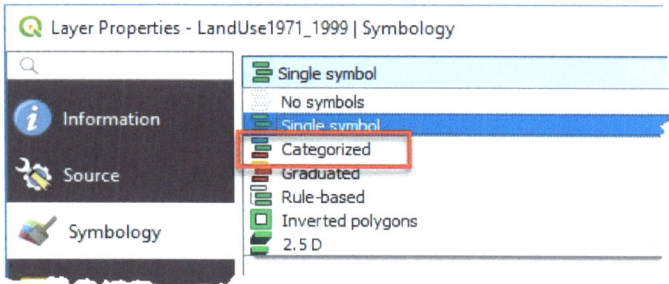

Figure 62: Begin by choosing Categorized from the drop-down menu.

Okay, let's shade this map. Choose Categorized from the drop-down menu (Figure 62).

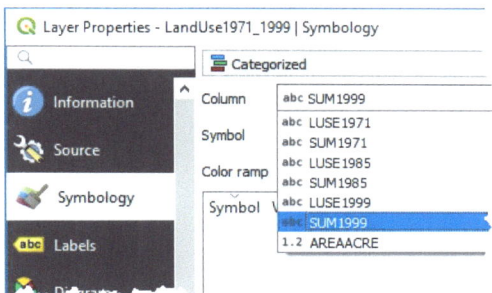

Figure 63: Now choose to map the contents of the SUM1999 column.

Then in the new dialog box that appears, choose to Shade by the SUM1999 column (Figure 63). Remember, these are the summary categories that I created for the year 1999.

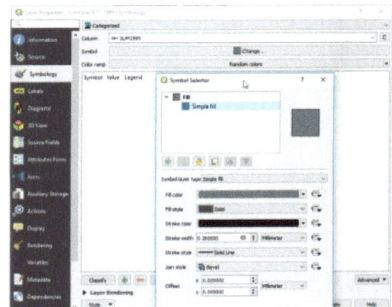 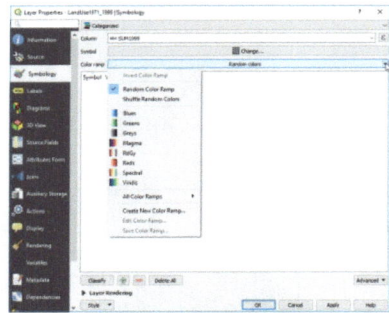

Figure 64: Ensure the fill style is solid.   *Figure 65: Choose Random Colours for the colour ramp.*

Ensure the fill style is solid (Figure 64). Choose the "Random Colours" colour ramp (Figure 65). Click on the Classify button and the map will shade.

I want you to experiment with the Color Ramps. Just choose a different ramp from the drop-down list and then click Apply button to apply it to the map.

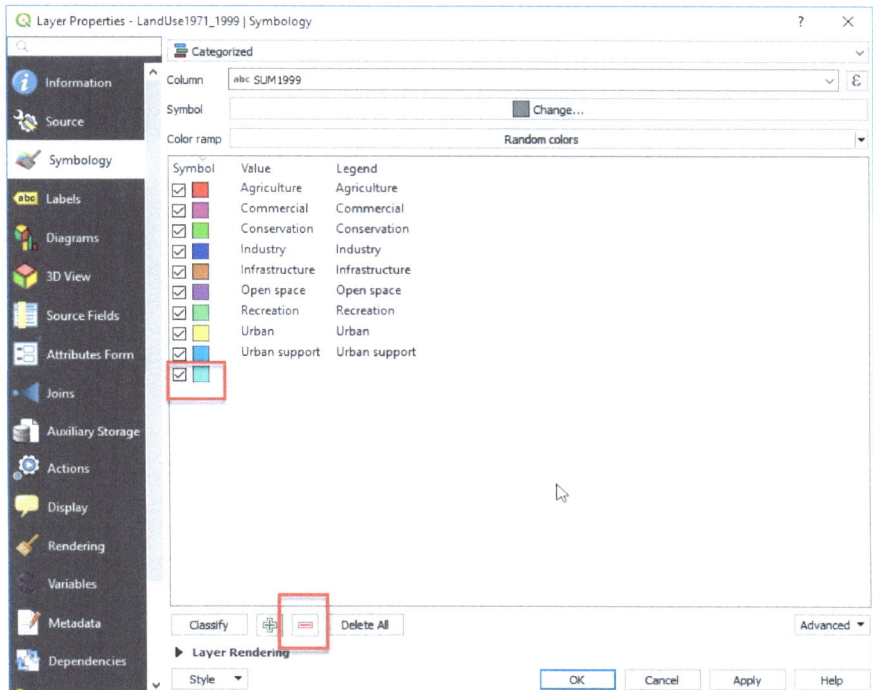

Figure 66: Remove the last legend category.

Select the last empty category in the legend. Use the minus button to delete it because we're not going to need that (Figure 66).

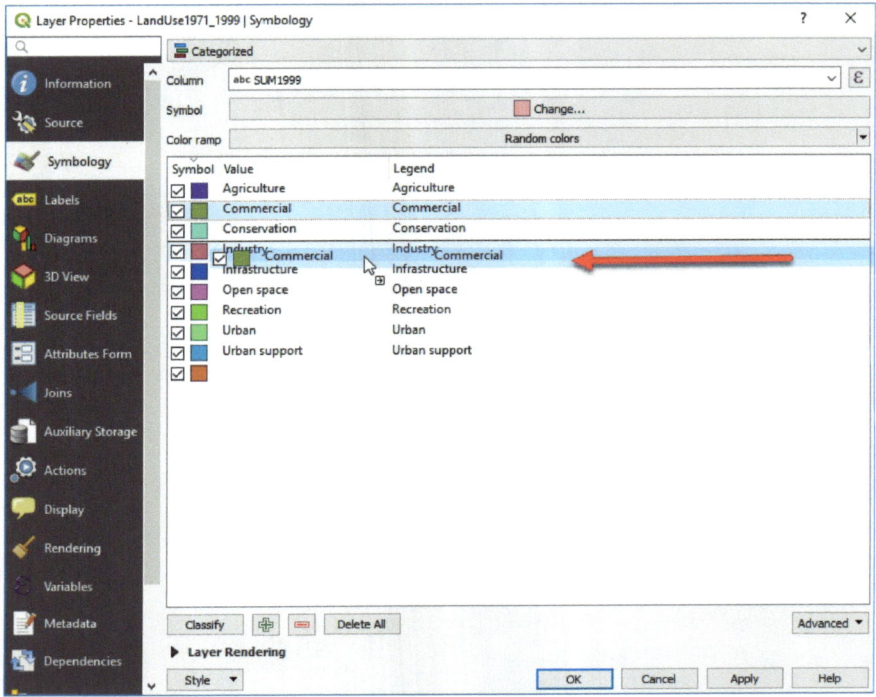

Figure 67: You can drag-and-drop similar categories next to each other.

Next, lets make the legend a bit more logical. Drag and drop similar categories next to each other (Figure 67). Click on a category, hold your left mouse button down, then drag-and-drop the category into a new place. The order I've got is Conservation, Open Space, Agriculture, Recreation, Infrastructure, Commercial, Industry, Urban and then Urban Support.

Figure 68: Enter the RGB values for a category's color in this dialog.

Now let's change the shading to a more attractive colour scheme. Here's an example I prepared earlier. Open the LandUse1999 Project file if you want to see it mapped (Figure 68). I've used…

- greens for open space related,
- pinks for commercial related,
- oranges for urban related.

Table 2: The Red, Blue, Green (RGB) values that I used for shading the categories in Figure 68.

Class	Red	Green	Blue
Conservation	60	138	71
Open space	0	170	27
Agriculture	0	255	127
Recreation	170	255	127
Infrastructure	255	85	127
Commercial	255	85	255
Industry	255	170	255
Urban	255	220	185
Urban support	255	170	0

Our map will look a whole lot better if we get rid of the black borders around each polygon. Click on the Stroke colour and shade the border the same colour as the polygon (Figure 68). It is tempting to select No Pen, but the borders show as white if you do.

This colour scheme was inspired by a GIS planning standard that we have in Victoria, Australia. But if you type into Google something along the lines of "GIS mapping standards", a whole bunch of standards will come up. Another search term for Google would be "map colour standards". Figure 70 is the standard that inspired the colour scheme I used in this tutorial.

Table 2 shows how that inspiration translated into the colour scheme we're using in this tutorial. It specifies colours as Red, Green, Blue (RGB) values. RGB is easy to reproduce in QGIS.

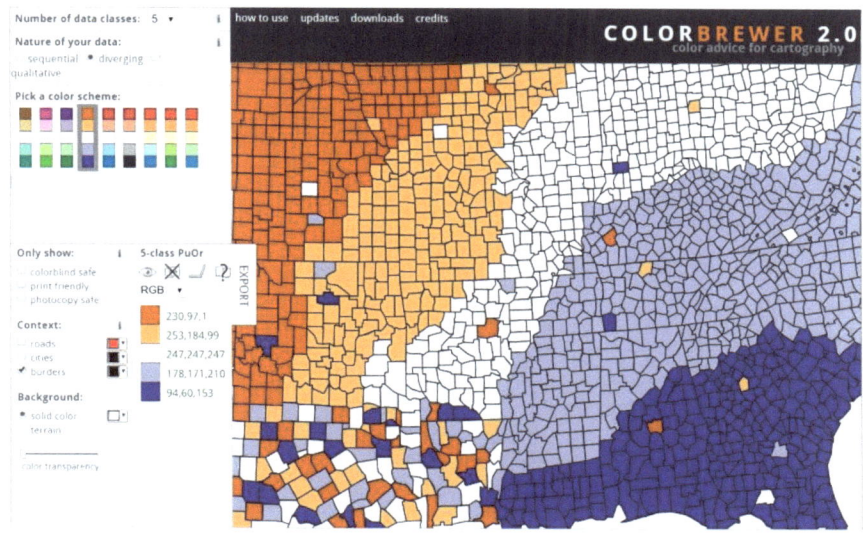

Figure 69: An example of a colour scheme from colorbrewer2.org.

Another option is ColorBrewer2.org (Figure 69). This site has a whole bunch of different colour schemes. Select the number of classes that you want, pick a colour scheme, and it will display in the sample map. The RGB values are displayed. And if you run your mouse over the ColorBrewer map, the codes for the different colors come up in a box.

The Colorblind Safe map ensures that even colorblind people can read the maps you produce. The ColorBrewer site is truly worth having a play with.

APPENDIX E
ZONE COLOUR VALUES

CODE	ZONE DESCRIPTION	MapInfo RGB Value
R1Z	Residential 1 Zone	255, 209, 204
R2Z	Residential 2 Zone	255, 181, 207
R3Z	Residential 3 Zone	255, 153, 204
LDRZ	Low Density Residential Zone	255, 166, 153
MUZ	Mixed Use Zone	217, 77, 77
TZ	Township Zone	255, 102, 153
B1Z	Business 1 Zone	240, 217, 250
B2Z	Business 2 Zone	224, 181, 242
B3Z	Business 3 Zone	194, 140, 178
B4Z	Business 4 Zone	153, 89, 166
B5Z	Business 5 Zone	128, 51, 140
IN1Z	Industrial 1 Zone	240, 176, 130
IN2Z	Industrial 2 Zone	204, 166, 128
IN3Z	Industrial 3 Zone	191, 89, 26
RLZ	Rural Living Zone	255, 204, 153
GWZ	Green Wedge Zone	204, 204, 153
GWZA	Green Wedge A Zone	204, 224, 153
RCZ	Rural Conservation Zone	204, 204, 0
RAZ	Rural Activity Zone	0, 204, 153
FZ	Farming Zone	222, 255, 237
UFZ	Urban Floodway Zone	153, 227, 255
SUZ	Special Use Zone * (plus No.)	222, 250, 138
CDZ	Comprehensive Development Zone* (plus no.)	5, 189, 194
CCZ	Capital City Zone* (plus no.)	25, 255, 235
DZ	Docklands Zone* (plus no.)	89, 89, 252
PDZ	Priority Development Zone* (plus no.)	255, 153, 255
UGZ	Urban Growth Zone* (plus no.)	238, 180, 180
ACZ	Activity Centre Zone* (plus no.)	153, 204, 204
PUZ1	Public use Zone - Service and Utility	255, 255, 153
PUZ2	Public use Zone - Education	255, 255, 153
PUZ3	Public use Zone – Health and Community	255, 255, 153
PUZ4	Public use Zone - Transport	227, 227, 227
PUZ5	Public use Zone – Cemetery/Crematorium	255, 255, 153
PUZ6	Public use Zone – Local Government	255, 255, 153
PUZ7	Public use Zone – Other Public Use	255, 255, 153
PPRZ	Public Park and Recreation Zone	161, 219, 178
PCRZ	Public Conservation and Resource Zone	97, 204, 38
RDZ1	Road Zone – Category 1	240, 10, 176
RDZ2	Road Zone – Category 2	255, 176, 0

Figure 70: An example of a metadata document with a GIS color scheme. There are many of these types of documents on the web. PRODUCT DESCRIPTION, VICMAP PLANNING, Information Services Branch, Department of *Sustainability* and Environment, Document version 2, Using Data Model Version 1.3, November 2012.

How to Validate Your Map

Maps used to be expensive beasts to create, and so there was lots of peer scrutiny along the road to production. A well presented map is a cartographers tool to tell a story in a convincing and authoritative way. GIS has de-skilled the traditional art of cartography. Too often unqualified people produce maps without considering the quality of the information they're using. And because these maps look so good, their audience takes them at face-value too readily.

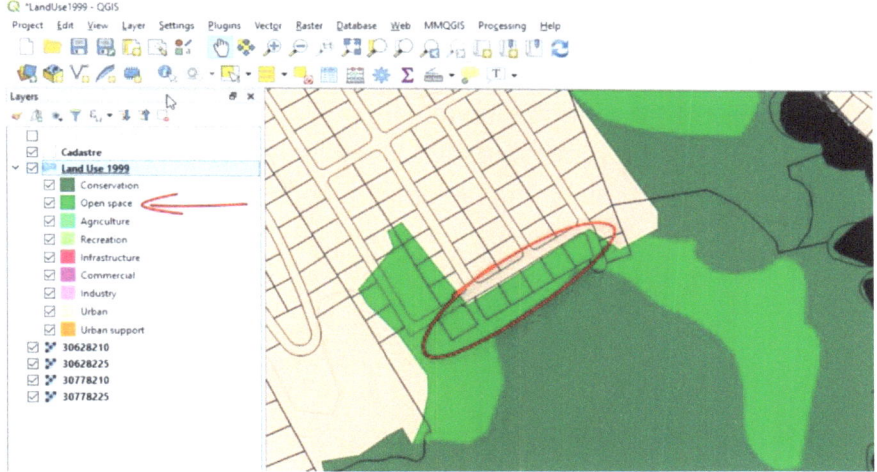

Figure 71: I quickly found this error in the Land Use map – small parcels of land in an area interpreted as being Open Space.

I'm big on validating maps. It took me less than a minute to find an error in our Land Use map (Figure 71). I want to show you how I went about finding that.

If you haven't already got the shaded 1999 land use map on-screen, open the LandUse1999 project...

- Add the cadastre and make it transparent with black lines.
- Add the land use map and make it transparent with pink lines. Pink is a great colour to use for this sort of work.

- Add the air photos too. Drag and drop these layers so the transparent land use and cadastre layers are on top.

Zoom into the area shown in Figure 71 and enable the cadastre. This green area has an Open Space category in our Land Use map, but yet the cadastre shows housing lots on it. Lets investigate this further. The metadata for our datasets tells me…

- The Land Use map was created using aerial photography from 1999.
- The cadastre was current in 2010 (11 years later).
- The air photography is from 2009 (10 years later).

This means that the Land Use interpretation is at least 10 years older than the other two datasets. So we need to be careful not to take the Land Use map at face value.

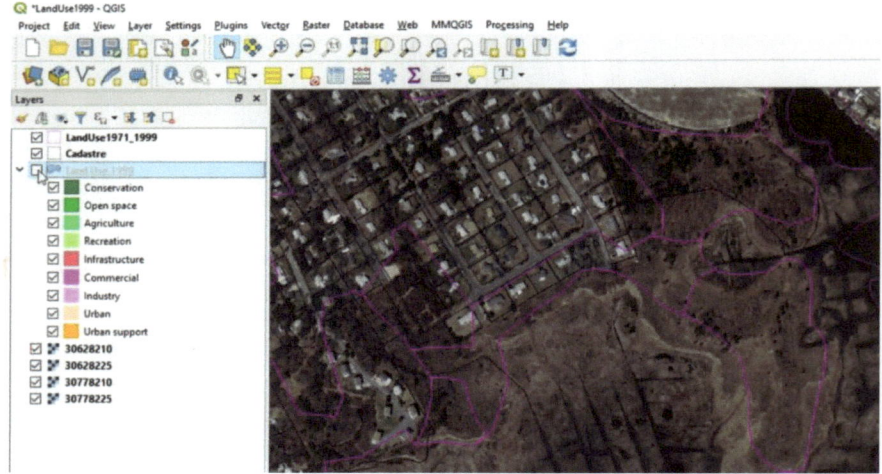

Figure 72: With the air photo displayed, you can see that the area interpreted as being Open Space has buildings in it.

Enable the aerial photography, move the two transparent maps to the top and disable the shaded Land Use map. We can now see the Land Use map and the cadastral map overlaying the aerial photography.

Let's zoom in a little bit closer (Figure 72). Obviously there's been some building going on within this open space. Maybe the land has been rezoned? Maybe it was already rezoned when the land use map was interpreted? So many questions!

However, the concept I want you to grasp is that you should not readily accept all maps at face value. Have a look at the metadata (the data that describes your GIS map). Try to understand whether a map has the potential to be out of date, and if so, use other mapping to help you understand whether or not there's data quality issues in the map. We were able to use the cadastre and aerial photography to confirm our suspicions. Other times you might use google maps.

This is a very big area of discussion. Books have been written about it. Data quality, or what I like to term as "Traps With Maps" is a constant theme through much of my teaching.

How to create a Time Series map

In this final part of the tutorial, I want to show you a technique for doing visual analysis of Land Use change. There are undoubtedly more sophisticated techniques for doing this. Techniques that would produce tables of quantitative data.

I mostly want to show you this visual technique for time-series analysis so you can to start getting your minds turning over with some of the possibilities for GIS analysis.

Now to the exercise…

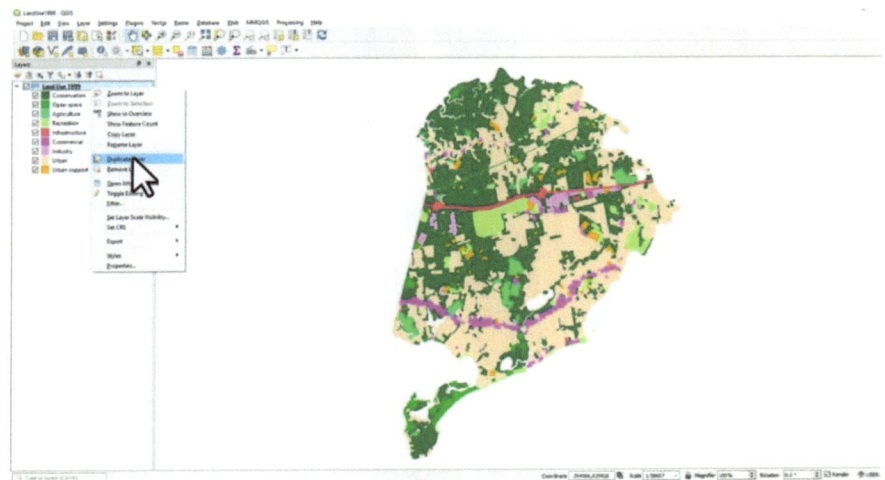

Figure 73: Duplicate the Land Use 1999 map twice.

Open the LandUse1999 project. Right click on the Land Use 1999 layer. Duplicate the map layer twice (Figure 73). This will allow us to thematically map the three time periods.

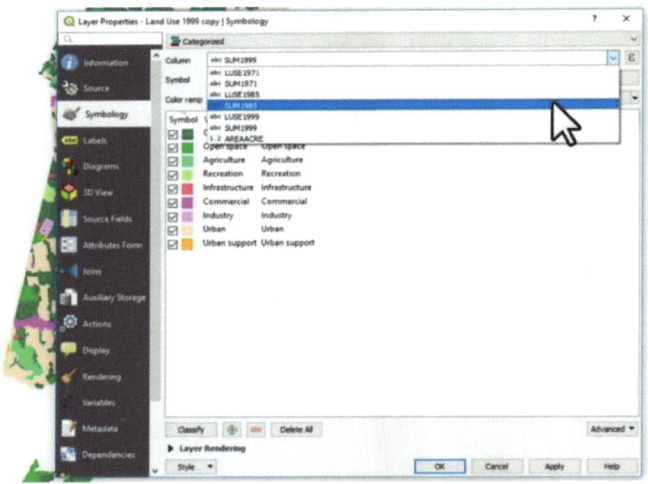

Figure 74: Change the column being mapped from SUM1995 to SUM1985.

Right-click on one of the Land Use 1999 maps. Go to Properties. Go to the Symbology tab. In the drop down list, change the SUM1999 column to be the SUM1985 column. Go to the Source tab and change the Layer Name to be Land Use 1985. Click Apply.

For a different map, repeat what we just did for the 1985 time period, but this time for 1971.

We now have three time periods thematically mapped.

If you need to, drag and drop the layers to reorder them so that 1971 is on the bottom, 1985 is in the middle and 1999 is on the top.

What's really important about this is that we were able to copy one map, and by simply changing the data column, we were able to apply our shadings and legend categories to a series of maps. This is not just time saving. It also makes our three maps consistent, and easy for an audience to understand.

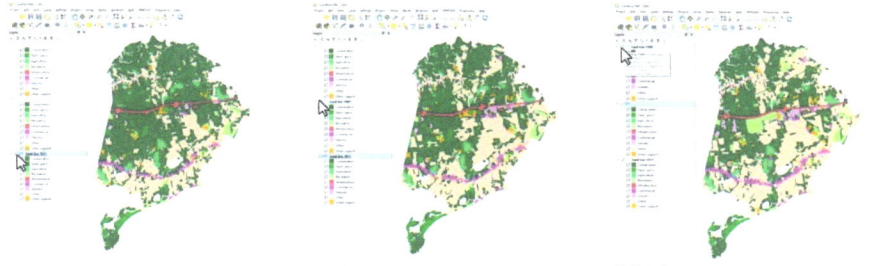

Figure 75: Land Use 1971.   *Figure 76: Land Use 1985.*   *Figure 77: Land Use 1999.*

Look closely at Figure 75, Figure 76 and Figure 77. There's a difference between...

- 1971 and 1985.
- 1985 and 1999.
- 1971 and 1999.

Notice how much the urban areas have expanded over that period. Imagine adding agricultural quality maps to this series. You'd gain insights as to whether the expansion of urban areas has the potential to undermine the agricultural sustainability of Yarmouth. See how the addition of seemingly unrelated GIS maps has the potential to increase the value of your GIS?

This same time series technique could be used for other types of studies and other fields of interests. You could map change in population density through time. Change in vegetation quality or habitat quality for critters. Epidemiological maps could show changes in health outcomes through time.

Play with this. Perhaps thematically map the detailed Land Use categories, and look at the change in those through time.

CONCLUSION

Well, that's it. I imagine that many of you will have focused on the practical parts of this tutorial. Be sure to take the time to look at the more theoretical parts too. Some of the Udemy videos are important to your understanding of how GIS works. For example…

- I reckon that John Snow's epidemiological study from the 1850s is truly amazing. Well before GIS, he used Spatial Analysis techniques to track down the source of a cholera outbreak.

- Then there's the Ian McHarg story. He was a planner who pioneered the routine use of map overlay techniques, and is regarded by many as being the father of GIS. I also touch on this in the Map Overlay section on page 12.

- Of course, there a video about points, lines and polygons too. I also talk about them in the Spatial Data (Maps) section on page 7.

- There's also plenty of GIS project examples from both my research and my consultancy life.

Please leave a review so you can help others learn how they can benefit from this book and help me learn how I can better serve my readers.

I hope you found this tutorial useful. Perhaps we'll meet again in one of my others?

IAN ALLAN

Thank you and take care!

Ian

COUPON FOR THE COMPANION VIDEO COURSE

Enrolment in the Udemy tutorial associated with this book is FREE.

GIS for Beginners #1: QGIS 3 Orientation: Learn to use QGIS 3. Navigate the interface. Create a shaded Thematic Map. Learn GIS basics and geospatial analysis.

Due to recent changes in Udemy policies (free coupons expire after 3 days) you will need to…

- email support@gis-university.com and request a free coupon from me.
- Please attach a copy/screen capture of your receipt to your request.
- Please use the subject heading "AMAZON FREE COUPON PLEASE"

I will return email a coupon. You will have three days to redeem the coupon.

Here's links to my other QGIS tutorials. Feel free to email request a discount coupon for these (support@gis-university.com). The coupon I send you will be for the lowest price Udemy will allow me to at the time (usually somewhere between $10 - $15 USD).

GIS for Beginners #2: Georeference & Digitize in QGIS3: Learn Georeferencing and Digitizing using Equipment Every Office has in this QGIS tutorial.

GIS for Beginners #3: Spatial Analysis using QGIS: Learn the Bare Essentials of Spatial Analysis - map overlay, spatial data query and buffering in GIS

GIS for Beginners #4: Learn Geocoding in QGIS 3: Learn Geocoding in QGIS 3. Geocode address data from spreadsheets. Geocode to Google, street lines and Address Points

GLOSSARY OF TERMS

Attribute data	Data that relates to a map object. For example, two attributes of a dot on a GIS map might be that… 1. It's a fence post 2. It's made of wood.
Categorical data	Sometimes called Data Classes. Data that can be expressed as groups. For example Land Use (rural, urban, etc), Vegetation Type (forest, grassland, etc), Habitat Quality (high, medium, low)
Clip board	Standard Windows functionality that allows you to highlight, then copy and paste text and pictures from one computer program (eg. a text editor) to another (eg. QGIS).
Column	The vertical collection of cells in a table. In a table, a column normally has a title which is its reference (for example "DATE"). Data follow underneath the heading (for example, October 17…). *See also, field*
Cross tabulation	The joint distribution of two variables. For example, a cross tabulation of "full time employees" and "teenagers" would reveal all those teenagers who are employed full time.

Database	A collection of tables that are used to describe the project you are working on. In a well designed database, each table will contain information about only one aspect of the project. It is very important that information is stored only once within a database, otherwise you can have problems maintaining data validity and integrity. For example, if a someone's address was stored in two database files and only one file was updated when the person moved house, how would you know which address was correct?.
Digitize	*See Digitizing tablet*
Digitizing tablet	A sophisticated electronic tablet on which you attach maps with sticky-tape and trace features with a *puck*. The puck interacts with a dense grid of wires inside the tablet that detect its position and the map features are digitized into the GIS. Later these features get interrogated by sophisticated software that checks its geographical integrity and then turns linework into enclosed polygons. These are not so common these days. Often you can get by by on-screen digitizing. I show you how to do this in my *GIS for Beginners #2: Georeference & Digitize in QGIS* tutorial
DPI	Dots Per Inch. This is scanner speak. The higher the number the more detailed the scan. When I'm georeferencing maps to digitize from, I usually scan them at 300 DPI.
Dynamic map	A map that changes in response to new information. It is always up-to-date. Online weather maps are a good example. Contrasts with a Static map.

Field	A reference to a column of data in a database. Imagine an Excel spreadsheet with a column called "date" and you're some ways to understanding the concept of a Field. However, in a database a field is like a column in an Excel spreadsheet on steroids. Database fields can be referenced by computer programs and GIS queries.
File	*See Table*
Geocode	A text description that can be related to a geographic object (eg. zip codes, addresses, census district names, and local government areas). Imagine you... 1. Have a 300 row spreadsheet. Each row represents a member of a local club. One of the columns contains each member's address. 2. You have a GIS map representing all local addresses You could create a map of your 300 club members by matching the address in each row of your spreadsheet to the GIS map. I show you how to do this in my *GIS for Beginners #4: Learn Geocoding in QGIS 3* tutorial.
Geographical Information System (GIS)	A computer based system for displaying, manipulating and analysing map based information.
GIS	See Geographical Information System.

Ground control point (GCP)	When you're geo-referencing a raster file (ge. a scanned map or an air photo), the scan is in a coordinate system that relates only to itself (ie. row and column numbers), and not to anywhere on the earth (eg. latitude and longitude). In order to relate your scan to an earth position you need to find places on your scan that you can also find in a GIS map. When you digitize these places they become known as *Ground Control Points (GCPs)*. When you've found enough GCPs you run a QGIS program to rectify the image. That turns your scan into an image that can underlay your GIS maps. I show you how to do that in my *GIS for Beginners #2: Georeference & Digitize in QGIS* tutorial.
Key field	A column in a table that contains a unique identifier that allows a row to be related with a row in another table. Customer Number and Student Number are examples. (see also *Field*). In the example below because the customer numbers in both tables are identical it is possible to print a delivery docket that ensures a Large Refrigerator gets delivered to 10 Smith Street.

Customer Number	Sale Item
1234	Large Refrigerator
1235	Small Refrigerator

Sales

Customer Number	Address
1234	10 Smith Street
1235	18 Jones Street

Customers

Line	A GIS object defined by two X and Y coordinates. For example, a section of road.
Map layer	A map of a single theme such as cadastre, roads, water features, etc. A GIS maps is created by combining multiple layers. In QGIS, maps can only contain one data type (ie. points, lines, polygons, polylines).

Map object	See object
Mental map	A map that is produced based on someone's understanding of an area. For example, a farmer's mental map of their land will relate to it's productivity and ease of management. An environmentalist's view of the same tract of land might relate to the quality of its wildlife habitat.
Metadata	Information about information. For example, in a census what does employed mean? Working >10, >20, >30 hours each week? Most data custodians have metadata describing their datasets. Often this can be found on the internet or in a library. If all else fails, contact the custodian by email or telephone!
Node	*See Vertex*
Object	This most often refers to something that is mapped such as a tree, track or paddock. For an explanation of the four GIS object types, see *Points, Lines, Polylines* and *Polygons*.
Point	A location defined by an X and Y coordinate. For example, a power pole or fence post.
Polygon	An area defined by three or more X and Y coordinates. The final coordinate is identical to the first coordinate. For example, a sports oval or a paddock.
Polyline	A location defined by two or more X and Y coordinates. For example, a power line or a fence.

Post Processing	Computer software is used to compare GPS collection to that of a base-station GPS in order to gain centimeter accuracies
Row	The horizontal collection of cells within a table. That part of a table that contains data. In a GIS, each row usually contains information about a map object.
Segment	This is the line between two vertices/nodes.
Select / Selection	You can select a map object by clicking on it with your mouse either on a map, or within a table. Either method will make a selection in both your GIS map and the attached. When you make a selection, you can choose to do things only to the items you've selected. For example, change its color on the map, or change a value in a column.
Sort	A process by which the values in a data column are ordered either alphabetically or numerically from lowest to highest. In a GIS, this is done using Structured Query Language (SQL).
Spatial	Anything that relates to "space" and can be mapped.
Static map	A map that doesn't change once its been produced. It remains fixed at the point of time in which it was produced. Printed maps are static maps. Contrasts with a Dynamic map.
Structured Query Language (SQL).	The language that a GIS uses to summarize data.

Table	A method of storing data. A table has columns with headings that can be referred to by a GIS, and rows containing data that are "used" by a GIS. A table is normally confined to information about one topic.

Item	Date	Crop
PicnicArea	Oct 17 2010	Sprayed
PicnicArea	Nov 17 2010	Mowed
PicnicArea	Dec 17 2011	Fertilized

Field identifier at top, *Column* on right, *Row* below.

See *field* |
| Temporal | Anything that relates to time. If we have two maps of the same theme over the same area that have been created at different times then we can map "change". For example, comparing a land use interpretation through time.

I show you how to do this in my *GIS for Beginners #1: QGIS 3.4 Orientation* tutorial. |
| Text | In GIS this means data that are stored in ASCII format. ASCII format data can be read by text editors and can also be read by QGIS. Variations that you're likely to come across are Comma Separated Value text (.csv), straight text (.txt) and Tab Separated text (normally you check a "Tab" box when importing this).

Word processor files and database files are NOT text data. |
Thematic map	A map about a theme. A theme might be demographic (total population in each postcode), environmental (habitat quality in each postcode), economic (total number of manufacturing plants in each postcode), etc. As you can see, any area can be mapped for numerous themes.
Vertex	When you're on-screen digitizing, each mouse click creates a vertex (sometimes called a node). See also segment.
Windows Clipboard	See *Clipboard*

www.ingramcontent.com/pod-product-compliance
Lightning Source LLC
Chambersburg PA
CBHW071556220526
45469CB00003B/1034